CONCEPTUAL FRAMEWORKS IN GEOGRAPHY
GENERAL EDITOR: W. E. MARSDEN

The Atmospheric System

An Introduction to Meteorology and Climatology

Greg O'Hare B.Sc. Ph.D.
Department of Geography,
Derbyshire College of Higher Education

John Sweeney B.Sc. Ph.D.
Department of Geography,
St Patrick's College, Maynooth,
County Kildare, Ireland

Maps and diagrams drawn by Ann Rooke and Paul Ferguson

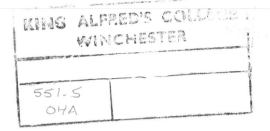
Acknowledgements

The authors and publishers wish to thank all those who gave their permission to reproduce copyright material in this book. Information regarding sources is given in the captions.

Cover illustration by Tim Smith.

Oliver & Boyd
Longman House
Burnt Mill
Harlow
Essex CM20 2JE

An Imprint of Longman Group UK Ltd

ISBN 0 05 003742 0

First published 1986
Seventh impression 1992

Set in 10/12pt Linotron Times Roman

Produced by Longman Singapore Publishers Pte Ltd
Printed in Singapore

The Publisher's policy is to use paper manufactured from sustainable forests.

Contents

Editor's Note

An encouraging feature in geographical education in recent years has been the convergence taking place of curriculum thinking and thinking at the academic frontiers of the subject. In both, stress has been laid on the necessity for conceptual approaches and the use of information as a means to an end rather than as an end in itself.

The central purpose of this series is to bear witness to this convergence. In each text the *key ideas* are identified, chapter by chapter. These ideas are in the form of propositions which, with their component concepts and the inter-relations between them, make up the conceptual frameworks of the subject. The key ideas provide criteria for selecting content for the teacher, and in cognitive terms help the student to retain what is important in each unit. Most of the key ideas are linked with assignments, designed to elicit evidence of achievement of basic understanding and ability to apply this understanding in new circumstances through engaging in problem-solving exercises.

While the series is not specifically geared to any particular 'A' level examination syllabus, indeed it is intended for use in geography courses in universities, polytechnics and in colleges of higher education as well as in the sixth form, it is intended to go some way towards meeting the needs of those students preparing for the more radical advanced geography syllabuses.

It is hoped that the texts contain the academic rigour to stretch the most able of such candidates, but at the same time provide a clear enough exposition of the basic ideas to provide intellectual stimulus and social and/or cultural relevance for those who will not be going on to study geography in higher education. To this end, a larger selection of assignments and readings is provided than perhaps could be used profitably by all students. The teacher is the best person to choose those which most nearly meet his or her students' needs.

W. E. Marsden
University of Liverpool.

Preface

Over the last few years, as the number of courses offered in meteorology and climatology has increased, and as the techniques available for investigating the workings of the atmosphere have improved, we have seen the need for a new textbook on the atmospheric system.

This text provides an up-to-date analysis of the elements and controls of climate and their inter-relationships. The atmosphere is introduced in Chapter 1, using systems-based theory. A consideration of atmospheric energy and motion, given in Chapters 2 and 3, serves as a foundation for later analyses concerned with atmospheric moisture, including precipitation processes and patterns (Chapters 4–6). Chapter 7 integrates previous chapters into an investigation of weather systems at the synoptic scale, and reinforces the idea of a complex atmosphere with its many interacting forces. The last part of the book departs from the normal presentation of the subject in that it does *not* offer an outline of climatic classification with descriptive accounts of the world's principal climatic regimes. Most textbooks which discuss in detail climatic classifications present schemes that are static and fixed systems. Such systems cannot therefore adequately elucidate our changing climate. Accordingly, Chapter 8 focuses on climatic change and emphasises once again the dynamic character of the earth's atmospheric environment.

No individual chapter of the book has been devoted exclusively to either microclimate or to applied climatology. Instead, numerous examples of the climate of a small area (for example, urban climates) and of climate–human interactions (for example, the effect of anthropogenic emissions of CO_2 on the thermal balance of the atmosphere) are discussed at appropriate points in the text.

Emphasis is given throughout to the temporal and spatial distribution of a wide range of climatic phenomena, the latter reflecting the book's essentially geographical viewpoint. The approach is furthermore non-mathematical to serve the needs of students in the arts and social sciences. At the same time, however, our approach is basically scientific, and some treatment of the physical principles of climate have been provided as necessary background information. To aid comprehension, a glossary of physical terms is given at the end of the book.

Finally, we would like to thank those who have given assistance during the preparation of the book.

Greg O'Hare
John Sweeney

1 Introduction

A. Systems

1. The systems approach

To understand the workings of the atmosphere, even in a rudimentary way, is both stimulating and demanding. In addition to this, however, a knowledge of how the atmosphere works is increasingly vital to humans, who are both users and managers of atmospheric resources. Such understanding is as yet rather elusive, owing to the colossal size and complexity of the mechanisms involved, and for this reason the construction of simple models is necessary. One such approach can be termed the systems approach.

Dismantling the functional linkages between phenomena with a view to analysing them is the essence of the systems-based approach. In the everyday world, we are only too familiar with this technique when we examine systems like a plumbing, or electrical, or heating system. In each case a structural or organisational entity can be isolated. In each case a flow or movement of energy and/or materials exists between the component parts of the system. In each case a driving force exists to power their operation.

2. Types of system

Three types of system may be distinguished (Figure 1.1).

Figure 1.1 Types of system

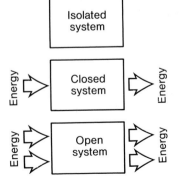

(a) Isolated system

In this there is no interaction between the system and its surroundings. No exchange of energy or materials exists across the boundary of the system. Such systems only exist in laboratory conditions and are unknown in the natural world.

(b) Closed system

In this there is a transfer of energy, but not of matter, between the system and its surroundings. Although this type of system is also rare, planet earth may conveniently be considered such a system. Solar and terrestrial radiation flows constitute the energy exchange. No material exchanges (with the relatively insignificant exception of meteorites) exist between the earth–atmosphere system and space.

(c)   Open system

In this system both energy and matter are transferred across the boundary. This is by far the most common situation, encompassing all environmental systems, including the atmospheric system itself. Understanding the pathways of these flows of energy and matter, and understanding the dynamic response of the atmospheric system to changes – both spatial and temporal – in them is thus the key to understanding the functioning of the atmosphere in all its complexity.

B.   The Atmosphere as an Energy System

1.   Internal composition

Air is a relatively uniform mixture of several chemical elements and compounds. Foremost among these are nitrogen (78%), oxygen (21%) and argon (1%), which account for all but a tiny amount of any volume of air that might be sampled. The remainder mostly consists of carbon dioxide, water vapour (especially in the lower atmosphere), ozone (especially in the upper atmosphere) and the noble gases – neon, helium, krypton and xenon – in decreasing amounts. In addition, compounds originating from biological or industrial processes such as methane, sulphur dioxide or nitrogen oxides also exist in trace amounts, as do solid and liquid particles of terrestrial (and occasionally extra-terrestrial) origin.

The importance of trace substances such as carbon dioxide, water vapour, ozone and particulate matter is much greater than their abundance might indicate. Water vapour, for instance, is the source of all clouds and hence precipitation; as will be seen in Chapter 4, the energy released when water vapour condenses is a major source of energy for the atmospheric system. Similarly, particulate matter and gases such as carbon dioxide and ozone are also important, exerting influences disproportionate to their abundance on flows of energy into and out of the atmospheric system.

2.   Vertical layering in the atmosphere

Investigations of the atmosphere enable it to be subdivided into a number of distinctive layers, according to temperature and zones of temperature change (Figure 1.2). Of most importance to humankind is the bottom layer, in which temperature decreases with increasing height. This layer is known as the *troposphere* and extends to about 12 km in altitude, where a temperature of −50 °C is fairly typical. The word *troposphere* is derived from a Greek word meaning overturning, and this is the key characteristic of this layer. Since the warmest levels are closest to the surface, the troposphere is relatively unstable. Almost all the features of weather and climate of human significance are contained in this layer, as is the bulk of the atmosphere's mass and almost all of its water vapour and dust particles.

Figure 1.2 Vertical temperature structure and principal layers of the atmosphere

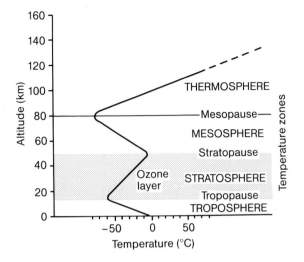

At the top of the troposphere temperature begins to rise with increasing height, reaching values above freezing point at altitudes of about 50 km. This is the *stratosphere*, and again the name betrays its main characteristic. In this case, since the warmest layers are aloft, stability is the main feature. The rising temperature curve is explained by the absorption of solar radiation by ozone molecules (O_3^-), which are more prevalent in this layer. The stratosphere is especially effective in screening out harmful ultraviolet rays from the sun, and is thus a vital shield for lifeforms at the surface. The possibility that humans could unintentionally damage the stratospheric ozone shield, by releasing harmful aerosol propellent gases which rise to this height, is a good reason why the atmospheric system needs to be understood. Another reason for investigating the atmosphere is the probable effect of a nuclear war on both the troposphere and the stratosphere. One apocalyptic scenario maintains that the billions of tonnes of dust and smoke released into the atmosphere upon nuclear exchange would virtually blot out the sun, plunging the earth into a prolonged 'nuclear winter'.

The stratosphere is overlain by a layer known as the *mesosphere*, within which temperature falls with increasing height. The boundary between the stratosphere and mesosphere, where this temperature decrease commences, is known as the *stratopause*. Above the mesosphere higher temperatures are again experienced, and this boundary is known as the *mesopause*. The air at this height, however, is extremely rarified and the molecules are widely separated as the vacuum conditions of space are approached. At a height of about 120 km, atmospheric pressure is only 0.001% of its sea-level value.

Key Ideas

A. Systems

1. A conceptual modelling approach is necessary to simplify the complex functioning of the atmosphere in order to aid our understanding of it.
2. Systems of various types can be identified, although they all share certain attributes:
   (a) an organisational entity, which can be identified;
   (b) flows of energy and material between their internal component parts;
   (c) an external driving force.
3. The earth–atmosphere, taken as a unit, may be considered as a closed system.
4. The atmosphere may be considered as an open system which responds to external and internal changes.

B. The Atmosphere as an Energy System

1. Although the atmosphere is predominantly composed of nitrogen and oxygen, important roles are played by trace substances such as carbon dioxide, water vapour, ozone and particulate matter (dust).
2. Layers within the atmosphere can be identified. Of these, the troposphere is of most human significance, although human impact may also be detectable in the stratosphere.

Additional Activities

1. Draw a diagram of the plumbing system in your home. Comment on:
   (a) whether it is isolated, closed or open;
   (b) its internal linkages;
   (c) its external driving force.
2. Give a reasoned account of the temperature changes that might be experienced during a manned balloon ascent to an altitude of 40 km.

2 Atmospheric Energy

A. Energy in the Atmospheric System

Energy in the atmosphere is responsible for all weather and climate. Atmospheric motion, for instance, including winds and storms, together with oceanic movement, is caused by the unequal heating of different parts of the planet. The nature and distribution of moisture in the air, whether in rain, snow or clouds, is likewise controlled by the atmosphere's energy conditions. There is also a close link between the energy and temperature characteristics of the atmosphere. This chapter examines the relationship between energy and temperature within the earth–atmosphere system. Attention is drawn to spatial and temporal variations in energy and temperature conditions as well as to changes induced by human action.

1. Energy input

(a) Solar radiation or insolation

All bodies emit energy to space in the form of electromagnetic waves called *radiation*. An important feature of these waves is their wavelength, that is, the distance from crest to crest of succeeding waves. Because the surface temperature of the sun is very high (about 6000 °C), the sun sends out energy to its surroundings mostly in the shorter wavelengths. Accordingly, *solar energy*, or *insolation*, is referred to as *short-wave radiation*.

Solar energy output provides by far the greatest source of energy input to the earth–atmosphere system. The quantity of input involved is of immense proportions. Insolation intercepted by the earth *each day* is sufficient to generate, or is equivalent to, 1000 large-scale depressions (low-pressure systems), 10 000 hurricanes or 100 million thunderstorms. On an annual basis, solar radiation receipt is more than 20 000 times the energy represented by fossil fuel-burning in 1985.

(b) Reflection and absorption of insolation

Insolation suffers a great deal of loss and dilution as it passes through the earth's atmosphere. About 27% (27 units) of the incoming solar radiation is reflected and scattered back to space by cloud and air molecules, and a

Figure 2.1 The components of the earth–atmosphere energy balance. All energy values are expressed as units of the original solar energy falling on the earth–atmosphere system (100 units)

(a) Solar radiation (short-wave)

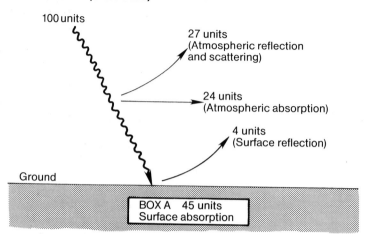

(b) Terrestrial and atmospheric radiation (long-wave)

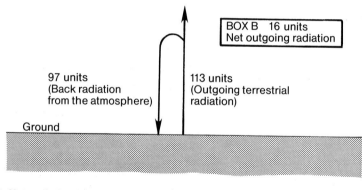

(c) Net radiation balance at the surface

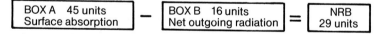

further 4% by the ground surface (see Figure 2.1a). The term *albedo* is used to denote the percentage or fraction of reflected short-wave radiation to total intercepted radiation. As the earth and atmosphere reflect 31% of the sun's incoming radiation, the average planetary albedo is 31% (denoted as 0.31). A further 24% of the incoming solar radiation is absorbed in the lower atmosphere by atmospheric gases and particles, especially water vapour, carbon dioxide, ozone and dust. Thus, of the insolation arriving at the outer layers of the atmosphere, less than one-half (about 45%) is actually absorbed by the ground surface. Despite this, the earth's atmosphere is often said to be relatively 'transparent' to short-wave solar radiation, because it tends to let a fairly large proportion (49%) pass through to the ground surface.

(c) *The role of solar radiation*

Since solar radiation is available continuously at the earth's surface, it is often referred to as a 'renewable' resource. As the principal source of primary energy input to the earth–atmosphere system, solar radiation performs a number of important functions.

(i) *Photosynthesis*. About one-half of the short-wave radiation emitted by the sun is composed of visible light. A small proportion of this visible light is fixed by plants to produce food by the process of *photosynthesis*. Because of this, solar radiation is crucial to the growth of plants, and their associated animals, within ecosystems.

(ii) *An energy supply for the world economy*. Solar radiation also provides an important energy source for human activities. Solar energy and photosynthesis in the geological past have produced the vegetation and animal life which is the source of present-day fossil fuels. Moreover, light and heat from the sun are employed in evaporating and raising water from the land and sea surface to the atmosphere. The return of this water as rain, and then river flow, to lower elevations is used as a source of hydro-electric power. In the future, as fossil fuels become scarcer (they are a non-renewable resource), the use of hydro-power as well as the direct use of solar power will probably become a major renewable source of energy for humankind.

(iii) *Planetary heating*. One of the principal functions of insolation is to supply heat for the ground and atmosphere. As indicated at the beginning of the chapter, this heat is responsible for all weather and climate, including atmospheric disturbances, clouds, rain and snow, and all the energy associated with them. Planetary heating is a fairly complicated process, however, and is examined in the next section.

2. Solar energy transfer and transformation

(a) *Terrestrial radiation*

Insolation absorbed at the earth's surface is converted into heat and warms the ground. The warm ground, in turn, radiates a good deal of energy as heat back to space, equivalent to 113% (113 units) of the original insolation reaching the earth–atmosphere system (see Figure 2.1*b*). Because of its relatively low surface temperature (on average 15 °C), the earth emits heat at fairly long wavelengths. All terrestrial radiation is therefore referred to as *long-wave* or *infra-red radiation*. An important characteristic of outgoing terrestrial long-wave radiation, however, is that, unlike solar short-wave radiation, it is largely absorbed by the atmosphere.

(b) *Atmospheric radiation*

Only a small part (16 units) of the infra-red radiation emitted by the earth's surface passes directly out into space. Most (97 units) is absorbed in the lower atmosphere by water vapour, water droplets in clouds and by carbon dioxide, and is subsequently returned as long-wave atmospheric radiation to the ground. In other words, the earth receives a second and major source of heating from the atmosphere. This back or *counter radiation* from the atmosphere to the ground is crucial in raising average surface temperatures. Without it, the mean surface temperature of the earth would fall by some 25 °C, making life virtually impossible.

(c) The greenhouse effect

The raising of surface temperatures as a result of counter radiation from the atmosphere is known as the *greenhouse effect*. As with the glass of a greenhouse, the atmosphere tends to allow solar short-wave radiation in, but partly stops terrestrial long-wave radiation from being lost again to space.

The role of water vapour and clouds in the atmosphere in the greenhouse effect is seen very well by comparing temperature conditions on clear and cloudy nights. On a cloudy, humid night, the cloud blanket absorbs heat from the ground and re-radiates it back again, maintaining relatively high night-time temperatures. On clear nights with little cloud, although water vapour and carbon dioxide absorb some energy, there is a greater escape of long-wave terrestrial radiation. As a result, the night-time temperature conditions are much colder (see Figure 2.12).

3. Energy balance sheets

(a) Net radiation balance

Quantitative details of the radiation processes already outlined are shown in Figure 2.1. It is possible to calculate from these an energy measure known as the *net radiation balance* at the surface of the earth. This is the difference between all incoming radiation at the earth's surface (that is, incoming solar energy and long-wave back radiation from the atmosphere) and all outgoing surface radiation to space (reflected solar radiation and outgoing long-wave terrestrial radiation). In other words, the net radiation balance at the surface is the difference between the absorbed solar radiation and the net outgoing long-wave radiation. If the original amount of insolation arriving at the outer layers of the atmosphere is taken as equivalent to 100 units of energy, then the absorbed insolation equals 45 units (box A). The net outgoing long-wave or infra-red radiation is equal to $113 - 97 = 16$ units (box B). Thus the net radiation balance is equivalent to $45 - 16 = 29$ of the original energy units (box A − box B).

(b) Heating of the ground and atmosphere

The net radiation balance at the earth's surface is employed in heating the ground, and subsequently, the air above. As indicated in Figure 2.2, about one-fifth, or 6 units, of the net radiation is used in heating the ground directly. A narrow zone of air in contact with the warm ground is heated subsequently by *conduction*. As warm air is less dense than the cooler air above, the former rises through the latter by *convection*. Winds

Figure 2.2 Surface to atmosphere heating: the role of the net radiation balance. One unit of energy is equivalent to 1% of the original solar energy falling on the earth–atmosphere system

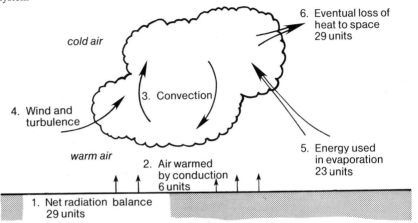

also bring air across the ground surface, mixing warm and cold air together by *turbulence*.

Most of the net radiation balance is used in the evaporation of water at, or near, the surface (23 units). To evaporate moisture a considerable amount of heat is required, approximating to about 590 calories for every gram so evaporated. This hidden or *latent heat of evaporation* is also transferred upward to the atmosphere by convection and turbulence. Latent heat is subsequently released back again to the atmosphere as actual, or sensible, heat during the cooling and condensation of this water vapour. Because solar radiation heats the ground directly, and the atmosphere only indirectly by the transference of heat from ground surface to atmosphere, there is normally a general decrease of temperature with altitude. This decrease is known as the *normal or environmental lapse rate*.

Finally, it is important to emphasise that, since the earth–atmosphere system is neither heating up nor cooling down, there must be a balance between incoming and outgoing radiation at the outer layers of the atmosphere. Consequently, the atmosphere must lose to space 29 units of energy to offset the net radiation balance.

ASSIGNMENTS

1. *Examine Figure 2.1.*
   (a) *How does solar radiation differ from terrestrial radiation?*
   (b) *Compare the ability of the atmosphere to absorb solar and terrestrial radiation.*
   (c) *What is the 'greenhouse effect' and why is it significant?*
   (d) *Define the net radiation balance and show how it may be calculated.*
   (e) *Analyse the stages and processes involved in heating the atmosphere.*
2. *What other functions are performed by solar radiation apart from atmospheric heating?*

15

B. Spatial Variations in Energy Characteristics

So far we have considered only the average planetary condition in the supply of insolation and the creation of a net radiation balance at the surface of the earth. In reality, there are considerable variations in space and time in the receipt and loss of radiation between different parts of the globe. This section examines these energy variables, focusing on the effects they have on prevailing temperature conditions.

1. Spatial variations in annual global insolation

The receipt of annual solar radiation at the earth's surface is primarily a function of latitude. This is because latitude determines the intensity of insolation at any particular place. Latitudinal patterns in solar radiation distribution are modified regionally and locally by the presence of cloud cover.

(a) Insolation intensity: the effect of latitude

Latitude determines the annual distribution of insolation by controlling the angle at which the sun's rays strike the surface. As shown in Figure 2.3, when the sun is low in the sky (high latitudes in winter), the sun's rays impinge on the earth's surface at an oblique angle. Such rays deliver less energy at the ground than vertical rays, when the sun is high in the sky (low latitudes). This is because:

1. the same amount of insolation is spread over a larger surface;
2. the same solar beam undergoes more severe atmospheric dilution by reflection, scattering and absorption in passing through a thicker layer of air.

Figure 2.3   The contrast of solar radiation intensity from vertical rays (high intensity) and oblique rays (low intensity)

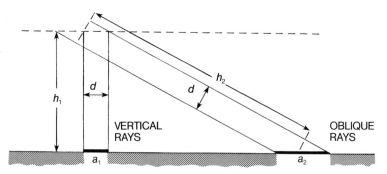

For the year as a whole, the angle at which the sun's rays strike the surface becomes more oblique polewards of the equator. As a result (assuming for the moment a cloud-free atmosphere), there is a general latitudinal decrease from the equator to the poles in the observed distribution of solar energy received at the earth's surface (see Figure 2.7).

(b) Regional patterns: cloud cover

Clouds are effective agents in reflecting and scattering solar radiation back to space. Thick cloud layers appear dark or black as seen from the ground surface, but white when observed from space because of their high solar reflectances. Table 2.1 gives estimates of the reflectance or *albedo value* for various types of material surface. Some of the highest reflectances are for particularly thick cloud formations, for instance, cumulonimbus and stratus cloud decks.

Table 2.1   Albedo values of various surface with some seasonal variations

	Albedo value (percentage)
A. Very high albedo surfaces	
1. Fresh snow	80–95
2. Thick cloud (average)	70–80
(a) Cumulonimbus (5 km +)	90–95
(b) Thick stratus (500 m)	60–70
B. High albedo surfaces	
1. Thin cloud (average)	25–40
2. Ice/sea	30–40
3. Saline deserts	25–50
4. Hot deserts	25–35
C. Moderate albedo surfaces	
1. Savanna (average)	15–25
2. Tundra (without snow cover)	15–20
Tundra (with snow cover)	80
3. Crops	15–25
4. Deciduous forest	15–20
D. Low albedo surfaces	
1. Green pasture and meadow (summer)	10–15
Green pasture and meadow (with snow cover)	70
Dry ploughed fields	10–15
2. Coniferous forest (summer)	10–15
Coniferous forest (with snow cover)	35–40
3. Urban areas	15
4. Dark soil	5–10
5. Oceans (average)	7–9
Oceans (sun near zenith)	3–5
Oceans (sun near horizon)	50–80

Figure 2.4   The observed average latitudinal distribution of annual solar radiation received at the earth's surface. (Source: Trewartha and Horn, 1980, p. 18)

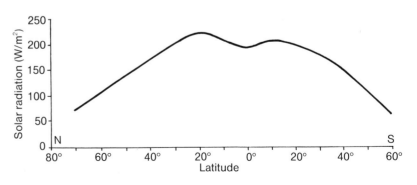

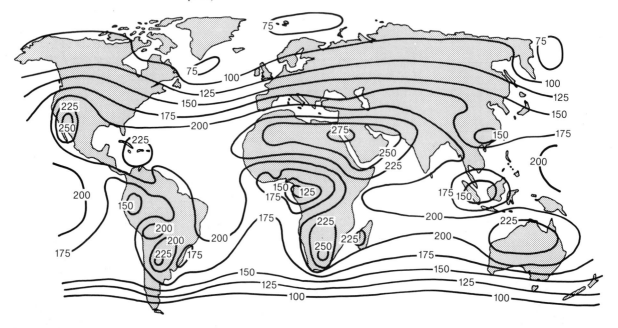

Figure 2.5 The average annual distribution of solar radiation. Insolation values as measured on a horizontal surface at ground level in W/m². (Source: Barry and Chorley, 1982, p. 19)

The high albedo values for most clouds have a marked regional effect on the distribution of received solar radiation at the earth's surface. Figures 2.4 and 2.5 reveal that the maximum annual insolation actually reaching the surface is found not at the equator, but rather in the cloud-free tropics and subtropics at about 20° N and 20° S. The lower annual insolation at the equator is due to the greater cloudiness of the equatorial zone (see Plates 2.1 and 2.2). At equivalent latitudes, insolation amounts in the southern hemisphere are lower than in the northern hemisphere. This is a result of the greater cloud cover of the southern hemisphere, where there is more ocean surface to provide water vapour for clouds.

2. Spatial variations in annual net radiation

The average net radiation balance at the earth's surface examined in section A of this chapter also varies spatially. Because insolation is such an important component of the net radiation balance, global annual patterns of the latter are determined by the strong latitudinal distribution of solar energy. Land and sea bodies, however, react differently to the interception, transformation and exchange of solar energy. As a result, contrasting patterns in net radiation amounts for land and sea are experienced at the regional level.

(a) Global scale: latitudinal and altitudinal contrasts

In response to the marked decrease of solar radiation from the equator to the poles, there is a sharp poleward decline in net radiation. This

Plate 2.1 Meteosat image of the disc of the earth taken on 7 July 1979. This example shows the Intertropical Convergence Zone (ITCZ) – a zone of uprising air – at X as a band of cloud just north of the equator. An extratropical cyclone with anticlockwise circulating air is illustrated off the west coast of Spain (Y). Albedo changes can also be observed. The high reflectance of cloud layers (e.g. X and Y) is indicated by their bright colour. On land, the bright sand of the Sahara (Z) is interrupted at intervals by darker rock outcrops (A) of the Ahaggar and Tibesti. The cultivated Nile delta (B) and valley contrast with the lighter surrounding desert sands. The grassland vegetation of southern Africa is noticeably darker in tone (C), but along the coast of Namibia the brighter land marks the return to desert (D). (Photograph: University of Dundee)

Plate 2.2   Meteosat image of the disc of the earth taken on 20 December 1984. In December the ITCZ has moved just south of the equator, indicated by a broken band of cloud over the Atlantic (A). Location X at this season comes under a sinking air stream, and a dry season prevails (see Plate 2.1). Station Z at about 25° N is again under a region of sinking air and experiences a desert climate. At the equator, Y is influenced by rising air throughout the year. As a result, clouds are present and rain falls in every month. (METEOSAT image supplied by the European Space Agency)

Figure 2.6   Variations in the average annual net radiation gain or loss by latitude: (*a*) by the earth's surface; (*b*) by the atmosphere; (*c*) by the earth's surface and atmosphere taken together. (Source: Sellers, 1965, p. 66)

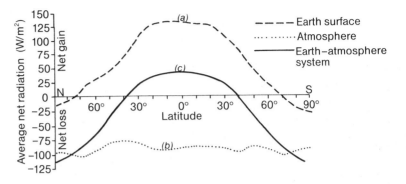

applies, as shown in Figure 2.6, to the whole earth–atmosphere system. Equatorwards of about 40° latitude, there is a *surplus of net radiation*. This means of course that more radiation (heat) from the sun is received by this zone than is lost by cooling through net outgoing long-wave radiation to space. Polewards of about 40° latitude, more radiation is lost to space than is ever directly received from the sun, that is, there is a *net radiation deficit* in this region.

This latitudinal pattern can be analysed as two separate components. When the earth's surface only is considered, a surplus of energy is found everywhere, except in polar areas where there is considerable reflectance and loss of insolation from cloud tops as well as from other high albedo surfaces such as extensive snow and ice-fields (see Table 2.1). It can also be seen that the atmosphere, taken by itself, loses more radiation than it gains, and this negative balance is fairly consistent with latitude. In view of these complex patterns in radiation receipt and loss (heating and cooling), three points stand out.

(i)   *Heat transfer: low to high latitudes.*   The negative net radiation balance for the earth and atmosphere *taken together*, which occurs polewards of 40° latitude, is compensated for by the transport of heat from low to high latitudes. If this were not the case, the low latitudes would grow continually warmer over time, and the high latitudes would become cooler. Atmospheric energy is transferred polewards by winds (about 80%) and by ocean currents (about 20%). In the middle latitudes, where heat transport is very large, mobile weather systems (for example, mid-latitude depressions or cyclones) are effective agents in this heat transport. Within the oceans, the great current systems (see Figure 3.12) carry warm water polewards and cold water equatorwards, and thus generate a net poleward transfer of heat.

(ii)   *Vertical heat transfer: ground to air.*   The negative radiation economy of the atmosphere itself is made good, or compensated, by the transfer of energy (heat) from the earth to the air, as outlined in section A of this chapter. Otherwise the atmosphere would constantly cool, while the surface of the earth would progressively heat up. It is important to realise that these differences in the heating and cooling of the earth–atmosphere system, both latitudinally and vertically, are the fundamental cause of all weather and climate.

(iii) *Temperature response.* In spite of the movement of heat from the lower to the higher latitudes, there is still a general decrease of surface temperature from equator to poles (see Table 2.2). Moreover, despite heat transport from ground to air, there is a general decrease of temperature from the ground surface through the atmosphere (see section C below).

Table 2.2   Mean annual temperature for different latitudes

Hemisphere	Mean annual temperature (°C)								
	0°	10°	20°	30°	40°	50°	60°	70°	80°
Northern	26	27	25	20	14	6	−1	−10	−17
Southern	26	25	23	18	12	6	0	−11	−20

(b)   *Regional patterns: land and sea*

While the net radiation balance and thus average temperatures at the earth's surface decrease from the equator to the poles, important regional variations in these two parameters are introduced by the disposition of the oceans and continents. This is because the sea differs from the land in the absorption, transfer and re-radiation of energy. As a result, it tends to heat up and cool down more slowly than the land. The most important reasons for surface temperature differences between land and sea include the following.

1.   The sun's rays penetrate water directly to a relatively greater depth than the ground, thus distributing their energy through a fairly deep layer. In contrast, the opacity of land concentrates the absorption of solar energy at the surface, producing rapid and intense heating.

2.   The fluid character of water permits vertical and horizontal mixing, thus minimising temperature contrasts. For instance, when water cools, it becomes more dense and heavy, and eventually sinks, allowing warmer water to rise to the surface.

3.   Water has a higher specific heat than the land. The *specific heat* of a substance is the amount of heat (expressed in calories) required to raise the temperature of that substance by 1°C. Whereas the specific heat of water is 1.0, the value for land is less than 0.5. Accordingly, to raise the temperature of a unit mass of water by 1°C requires more than twice as much energy as is needed to effect a similar temperature change in an equal mass of land.

4.   More solar energy goes into evaporating water over the oceans than over the land, and is thus not available for heating the surface and raising temperatures directly. Much of the latent heat employed in evaporating water over the oceans is transferred away from the water surface by wind to be released during condensation in the upper atmosphere, or over colder land and sea surfaces.

In view of these processes, continental land masses, especially in high latitudes, experience a wide range of temperature between summer and winter, and between day and night. On the other hand, oceanic areas show a more equable distribution of temperature, both on a seasonal and daily basis. These attributes of 'continentality' and 'oceanicity', together with variation in energy characteristics over time, are more usefully considered in the next section.

ASSIGNMENTS
1. *Consult Figures 2.4 and 2.5 and Plates 2.1 and 2.2.*
   (a) *Describe the pattern of annual average solar radiation shown in Figures 2.4 and 2.5.*
   (b) *Explain the observed distribution in terms of (i) latitude and (ii) cloud cover.*
2. (a) *Describe the energy relationships shown in Figure 2.6.*
   (b) *Explain the importance of these relationships in relation to: (i) the transport of heat from low to high latitudes; (ii) the transfer of heat from ground to air; (iii) the average global distribution of temperature from low to high latitudes and from ground to air.*

C. Temporal Variations in Energy Characteristics

There are large variations in the receipt of solar and net radiation at the earth's surface in time as well as in space. Important cyclical changes in energy patterns exist in relation to (i) the swing of the seasons and (ii) the passage of day and night. As a consequence, there are notable medium and short-term adjustments in the temperature condition of the lower atmosphere.

1. Medium-term seasonal adjustment

(a) Seasonal variation in solar radiation

If it is assumed for the moment that the atmosphere is cloud-free, it is possible to calculate the latitudinal distribution of insolation at sea level

Figure 2.7 Latitudinal distribution of solar radiation falling on a horizontal surface at ground level for the equinoxes and solstices. It is assumed that the atmosphere reflects 30% of the incoming solar radiation for vertical sun and cloudless skies. (Source: Trewartha and Horn, 1980, p. 18)

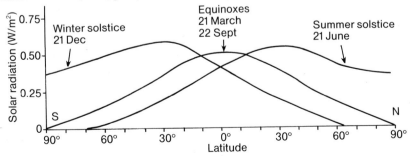

for the equinoxes and solstices (Figure 2.7). At the two equinoxes (21 March and 22 September), when the noon sun is directly overhead or vertical at the equator, solar radiation is evenly distributed in the northern and southern hemispheres with a maximum in equatorial latitudes and minimum values at the poles. During the solstices (21 June and 21 December), on the other hand, when the sun's noon rays are vertical over the Tropics of Cancer (23.5° N) and Capricorn (23.5° S) respectively, solar radiation distribution is very uneven between the two hemispheres. In each case, the 'summer' hemisphere receives between two and three times as much insolation as the 'winter' hemisphere.

(b) World distribution of January and July temperatures

The annual march of insolation between the summer and winter solstice creates very different patterns in hemispherical heating during January and July, as indicated in Figures 2.8 and 2.9. The solar or latitudinal control of temperature is evident not only in the temperature contrasts between the two hemispheres, but also in the east–west trend of the isotherms (or lines of equal temperature) in both cases.

The thermal effects of land and sea distribution, as well as that of the major ocean currents, also have an impact on global temperature distribution. This impact may be summarised as follows. First, there is a more pronounced migration and concentration of isotherms over land masses than over the oceans. Second, the annual range of temperature is greater in continental than in coastal locations. The annual range reaches an ab-

Figure 2.8   The world distribution of mean January temperatures (°C) at sea level. (Source: Barry and Chorley, 1982, p. 21)

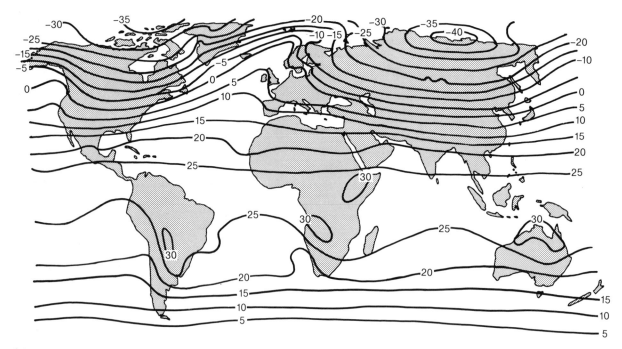

Figure 2.9 The world distribution of mean July temperatures (°C) at sea level. (Source: Barry and Chorley, 1982, p. 23)

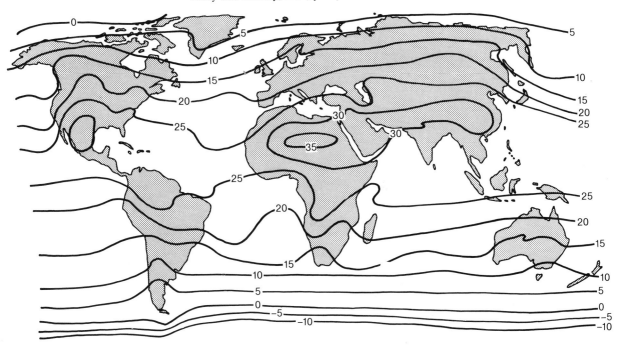

solute maximum of over 55°C in north-east Siberia. Third, the large heat storage of the oceans causes them to be warmer on average in winter, but colder in summer than land in the same latitude. Finally, the influence of the main ocean currents is evident especially in winter (for warm currents) and summer (for cold currents) (see Figure 3.12). The outstanding effect of the warm North Atlantic Drift in pushing isotherms polewards in the North Atlantic in January is clearly shown. So, too, is the pronounced equatorward displacement of isotherms along the coast of Peru and Chile in January (southern summer) by the cold Peruvian Current.

(c) Surface temperature anomalies

The significance of continentality and the influence of the ocean currents can be observed more clearly using the concept of the *temperature anomaly*. Thermal anomalies reflect not north–south (that is, latitudinal), but east–west (regional) changes in temperature. They are calculated by subtracting the mean (January or July) temperatures for all stations along a latitude from that of the mean (January or July) temperature for any individual station along the same parallel. The temperature deviation between a station and its parallel, whether positive or negative, is its anomalous temperature. As shown from Figure 2.10, the largest thermal anomalies are in the northern hemisphere, where continents and oceans alternate, and are seen to best effect in the January or winter period. At this season positive anomalies tend to be found over the oceans, whereas negative ones occur over the land masses. North-west Europe has the

25

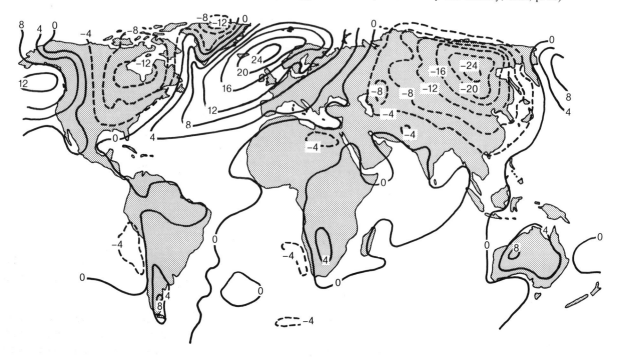

Figure 2.10 World temperature anomalies (°C) for January. Solid lines indicate positive, and broken lines show negative values. (Source: Barry and Chorley, 1982, p. 28)

highest positive temperature deviation on earth, where coastal Norwegian stations are over 20°C warmer than the mean for their latitude in January. In contrast, the extreme negative anomaly is found in north-east Siberia, which is over 26°C colder than the average for its parallel.

2. Short-term daily response

(a) The standard surface model

During most of the daytime, when the sun is high in the sky, incoming solar energy arrives faster at the ground surface than it can be dissipated as net outgoing terrestrial radiation. This occurs between the daytime hours A and B shown in Figure 2.11, and results in the accumulation of an energy surplus. As indicated, this energy surplus causes temperatures (T) to increase at the ground surface and thus in the air immediately above. During the early morning and late evening, when the sun is low in the sky, and at night, the rate of energy removal from the surface is greater than the rate of energy gain. As a result, the ground and the air above it undergo a reduction of temperature (that is, between B and C and D and A in Figure 2.11).

Such patterns in heating and cooling produce a daily march of temperature. The difference between the daily maximum and minimum surface air temperature is called the *daily range of temperature*.

Figure 2.11   A standard surface model of daily atmospheric heating and cooling in relation to radiation gains and losses under clear skies. (Source: Oke, 1978, p. 32)

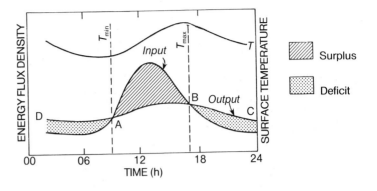

(b)   Modifications to the standard model

There are a number of conditions which produce modifications in the daily radiation–temperature sequence shown in Figure 2.11. The main factors include the following.

(i)   *Cloudy skies*.   Overcast cloudy days restrain the outflow of terrestrial energy and act to reduce nocturnal radiation cooling at the ground surface and in the air immediately above. As we have seen, clouds also reflect a large fraction of the incoming solar energy, reducing the daytime temperature increase of ground and atmosphere. Thus, although the effects of clouds on the *average* temperature over 24 hours may be quite small, they strongly suppress the daily solar control of temperature common under clear skies. Inspection of Figure 2.12 shows that a daily temperature range of 15 °C for a clear English summer day (not far from the average) can be superseded by a virtual collapse of the daily temperature cycle, when skies become overcast with intermittent drizzle.

Figure 2.12   The effects of cloudy skies on the daily radiation–temperature (°C) sequence at Reading University, 28 August–1 September 1975. A diurnal temperature range of 17°C on 29 August during clear skies was followed by suppression of the daily temperature cycle on 31 August, when the sky was overcast with intermittent drizzle. (Source: Riehl, 1978, p. 47)

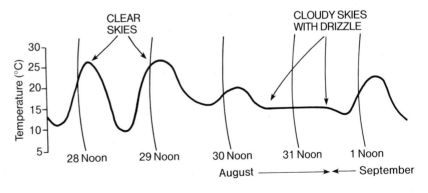

(ii) *Continental and maritime locations.* Small diurnal ranges are also found over oceans and over land with a maritime climate (for example, where winds blow from the sea). Since a water surface, for reasons outlined in section B above, tends to show little variation in temperature over a 24-hour period, neither does the air above it. In contrast, continental stations, especially those with clear skies in the subtropics, tend to show a larger diurnal range of temperature.

(iii) *Ground surface: snow cover.* Both day and night temperatures are lower in an area with a snow cover than one without. First, this is because snow is an excellent reflector (high albedo) of insolation, thus reducing net radiation at its surface. Second, snow is a very efficient insulator, because of the large amount of air it contains. It thus prevents heat reaching the surface of the snow from the ground below. Third, being a good emitter of long-wave terrestrial energy, heat is readily radiated at night from a snow cover, producing a cooling effect.

(c) Model of vertical temperature change

Until now we have considered diurnal temperature change in the air immediately above a warming (by day) and cooling (by night) land surface. Figure 2.13 shows the diurnal temperature response at various points upwards from this surface through the atmosphere. Two main phenomena are highlighted: the temperature lapse rate; and the temperature inversion.

Figure 2.13 Diurnal air temperature change with height under clear skies: (*a*) the temperature lapse rate; (*b*) the temperature inversion; (c_1 and c_2) the daily range of temperatures of different heights

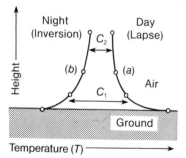

(i) *Temperature lapse rate.* During the day, temperatures normally decrease with distance away from the warmed ground surface, producing a *temperature lapse*. As explained on page 15, this is the normal temperature profile or condition for the whole atmosphere.

(ii) *Temperature inversion.* At night, especially under clear skies and calm weather, there can be a substantial loss of terrestrial infra-red radiation at the surface. This results in a marked cooling of both the ground surface and of the air immediately above (Figure 2.13). There is thus an increase of temperature with height in the lower atmosphere close to the ground. The occurrence of a negative lapse rate like this is known as a *ground-based temperature inversion*. These inversions are responsible for decreasing the daily range of temperature with altitude. Ideal conditions for ground surface inversions include: first, long nights – in winter – when there is a marked loss of net outgoing radiation; second, clear cloudless skies so that the loss of terrestrial radiation is rapid; third, relatively dry air which absorbs little outgoing radiation; fourth, calm stable air conditions so that turbulent mixing does not bring a mass of warm air from above to lower levels; and fifth, a snow cover which, because of its radiation, insulation and reflective qualities (mentioned earlier), cools the air immediately above much more effectively than a ground surface without a snow cover.

ASSIGNMENTS

1. (a) Describe the main features of world temperature distribution for January and July shown in Figures 2.8 and 2.9.
   (b) Account for the distribution patterns in terms of (i) the availability of solar energy (see Figures 2.6 and 2.7); (ii) variations in land and sea; and (iii) the arrangement of the ocean currents (see Figure 3.12).

2. (a) Define the term temperature anomaly.
   (b) Describe the distribution of temperature anomalies for January shown in Figure 2.10.
   (c) Compare the value of Figures 2.8 and 2.10 in demonstrating the influence of continentality and the major ocean currents on global temperature distribution.

3. (a) Describe and explain the relationships shown in Figure 2.11.
   (b) Examine the effect of cloudy skies, continental and maritime locations, and a snow cover in modifying these relationships.
   (c) In what ways will the daily march of temperature be affected by (i) thin, but clear atmospheres in high mountain areas; and (ii) a wet, as opposed to a dry, surface soil?

D. Human Impact on Atmospheric Energy

Human activities can alter the energy balance and therefore the thermal characteristics of an area in a great variety of ways. Table 2.3 shows a simple classification of the main results of human interference on the energy balance. Examples range from local, deliberate and beneficial alteration (e.g. the effects of a glass greenhouse) to unconscious and potentially adverse interference at the global level (e.g. the impact of carbon dioxide accumulations in the atmosphere).

A useful way of examining the effects of human disturbance on the energy balance a little further is to consider the mechanisms of interference. Humans are able to change inputs, transfers and outputs of radiation and thus can alter temperature patterns at different scales by (i) altering the nature and composition of the air, and (ii) changing the character of the earth's surface.

Table 2.3 A simple matrix with examples of human impact on the radiation balance

| Type of impact | Scale | | |
	Local	Regional	Global
Direct and deliberate	Warming within glass greenhouse	Use of carbon black dust to increase snow melt or fog dissipation	Climatic change through melting Arctic ice-caps, by pumping warm Pacific water across Bering Strait
Indirect and inadvertent	Urban heat island effect	Desertification may lead to climatic cooling	Atmospheric CO_2 warming effect: particulate cooling effect

1. Change in atmospheric composition

(a) Local and regional effects: the urban heat island

Anthropogenic alterations to the composition of the atmosphere are responsible for contributing to local and regional modifications of climate. As indicated in Table 2.4, many urban areas are 0.5–1.0°C warmer on an average annual basis and have winter minimum temperatures 1–3°C higher than their rural surroundings. The heat dome which hovers over the city has been termed the *urban heat island*. In Figure 2.14 we see a typical instance of the heat island during winter nights in the city of Dublin. A closed island-like system of isotherms, embracing warmer temperatures, separates the city from the general and cooler temperature field associated with the rural environment. The heat island effect can be explained, in part, by heat and pollution release within the urban area (see also Chapter 5, section D).

Table 2.4   Typical climatic changes caused by urbanisation

Type of change	Comparison with rural environs
Temperature	
Annual mean	0.5–1.0°C higher
Winter minima	1.0–3.0°C higher
Relative humidity	
Annual mean	6% lower
Winter	2% lower
Summer	8% lower
Dust particles	10 times more
Cloudiness	
Cloud cover	5–10% more
Fog, winter	100% more frequent
Fog, summer	30% more frequent
Radiation	
Total on horizontal surface	15–20% less
Ultraviolet, winter	30% less
Ultraviolet, summer	5% less
Wind speed	
Annual mean	20–30% lower
Extreme gusts	10–20% lower
Calms	5–20% more
Precipitation	
Amounts	5–10% more
Days with over 0.5 cm	10% more

Source: Matthews *et al.*, 1971, p. 168

(i)   *Heat release effect.*   Although solar radiation is still by far the most important energy source in human affairs, the residual heat release from fossil fuel combustion is comparable or larger than the solar input in some local urban areas. This is particularly the case in the winter period, when heat release from combustion is greater and solar radiation receipts lower than during the summer. The anthropogenic energy emission over the 60 km^2 of the Manhattan area of New York is almost four times the insolation falling on the area in winter. Even over sprawling Los Angeles (3500 km²), annual heat release from human activities now totals 13% of the solar radiation income.

Figure 2.14 The Dublin urban heat island for 20.00 to 1.00 GMT, 22 November 1983. (Isotherms in °C)

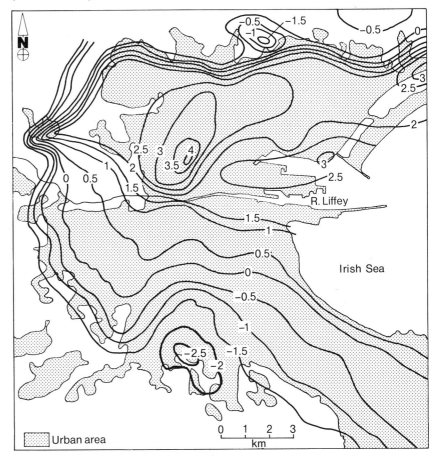

(ii) *Pollution effect.* The energy balance of a city may also be rather different from adjacent countryside (again, mainly in winter) because of pollution emission. The characteristic blanket of pollution, haze and cloud which hangs over many cities increases the atmospheric albedo, and reduces solar radiation at the urban surface (see Table 2.4). Nevertheless, a more important consequence of such pollution may be to return more outgoing long-wave radiation back to the urban surface, thus enhancing local temperatures.

(b) *Global effects: carbon dioxide and particulates*

On a global scale, potentially harmful modifications of the energy balance may result from: increases of atmospheric carbon dioxide and other 'greenhouse' gases; and/or changes in the quantity and character of particulate matter.

(i) *Carbon dioxide: the greenhouse heating effect.* Since pre-industrial times the concentration of carbon dioxide in the atmosphere has increased by 13–14%, from about 290 ppm (parts per million) in the 1870s to around 335 ppm at the present time. The major contributor to this increase is the release of carbon dioxide into the atmosphere from the combustion of fossil fuels. Over the past thirty years, fossil fuel combustion has expanded on average by over 4% per annum.

The increasing atmospheric concentration of carbon dioxide is of concern because of the 'greenhouse effect'. Carbon dioxide acts in some ways like the glass of a greenhouse. The gas permits short-wave radiation to pass through to the earth's surface to heat the ground and atmosphere, but intercepts some of the heat radiated upwards as long-wave radiation from the surface towards space. It then re-radiates long-wave energy back to the surface. With steadily increasing concentrations of carbon dioxide, the balance between incoming and outgoing radiation can be maintained only if temperatures at the surface and in the lower air increase.

It is possible to associate the post-industrial rise in atmospheric carbon dioxide concentration with temperature fluctuations in the northern hemisphere. As shown in Figure 2.15, actual temperatures in the northern hemisphere rose by 0.4–0.6 °C between 1880 and 1940, so it seems reasonable to implicate increasing carbon dioxide levels as a possible reason for this increase. It is difficult, however, to explain the 0.2–0.3 °C *drop* in temperature in the northern hemisphere between 1940 and 1965. Yet the apparent levelling off in the temperature in the northern hemisphere and the slight increase in temperature in the southern hemisphere since 1965 implies that the greenhouse effect may be becoming dominant once more.

A continuation of current trends in carbon dioxide emission (and this may or may not happen!) could increase average global carbon dioxide levels to around 600–650 ppm by the end of the twenty-first century. At this concentration, the expected carbon dioxide-induced global warming effect will equal 5–6 °C. Such speculative warming may result in the disappearance of the perennial Arctic ice, a partial disintegration of the West Antarctic ice-sheet and, as a consequence, a sea-level rise of 5–7 m.

Scenarios of a carbon dioxide-warmed world envisage significant effects on marine and land biota, agriculture and society. Many coastal areas, including the majority of the great cities of the world, would be seriously flooded by the predicted rise in sea level. Moreover, as predictions imply a progressive desiccation of the mid-latitudes around 40° N, they have alerted us to the possibilities of increasing drought in the main grain-growing areas of the world.

(ii) *Particulates: the dust-cloud cooling scenario.* The atmosphere carries, at any given time, billions of tonnes of many different kinds of particles, at various concentrations at different altitudes. Dust, soot and other particles enter the atmosphere from active volcanoes, forest fires, dust storms, sea spray (salt) and other natural sources. Humans also add particles by clearing land for agriculture, by inducing desertification, by urbanisation, and through industrial and automobile emissions.

A general atmospheric effect of particulates involves the scattering and reflection of solar radiation back to space. In this way particulates may reduce the amount of insolation reaching the earth's surface, and can thus induce a cooling effect. The huge quantity of particles poured into the high atmosphere from a major volcanic eruption can apparently cause regional and global cooling for several months and even years after the eruption (see section D of Chapter 8 below). With this effect in mind, and because there were few major eruptions between 1940 and 1970, it has been speculated that some of the drop in mean atmospheric temperature in the northern hemisphere between about 1940 and 1970 (see Figure 2.15) might be the result of increased atmospheric inputs of particulates from human activity. The recent apocalyptic idea that civilisation itself might be wiped out by even a modest nuclear exchange is based on the dust-cloud cooling response. Many eminent scientists in both the USA and the USSR now believe that a nuclear war would release such quantities of smoke and dust into the atmosphere that the sun's rays would be almost completely blotted out over the face of the earth for many months. Such a 'nuclear winter' would plunge the average temperature of the earth to below freezing point, making life and civilisation virtually impossible.

Nuclear war apart, it is not possible to be certain about the exact role played by present-day rates of anthropogenic particulate pollution on the temperature balance of the planet. This uncertainty is mainly the result of, first, a lack of reliable data on the amount of particulates of human origin added to the atmosphere. Estimates range, for instance, from 5% to 50% of the total atmospheric particulate loading. Second, there is a lack of knowledge of the complex interaction between short/long-wave radiation and atmospheric particulates. As already suggested, some particles, especially high-altitude naturally occurring ones (for example, fine volcanic dust), tend to reflect solar energy back to space, producing a cooling effect at the surface. Alternatively, larger particles of human origin concentrated in the lower atmosphere (for example, industrial smoke) might actually absorb such radiation and provoke a warming response. Moreover, outgoing long-wave terrestrial radiation may pass either through particulate concentrations or be reflected back to earth again.

On the other hand, particulate emissions could have a large indirect effect on the energy balance of the planet by influencing the amount and distribution, and thus the albedo value, of clouds. Particulates and aerosols are known to provide condensation nuclei (see page 80) and may thus be responsible for increasing cloud formation and cover.

2.   Earth surface change

(a)   *Local and regional effects: the urban heat island*

The urban heat island phenomenon examined above is also a function of the different physical surface character of urban areas compared with rural environments.

As shown in Table 2.1, urban surfaces have an albedo of about 15%; whereas evergreen forests have an albedo of 10–15%, deciduous forests 15–20%, and dry ploughed fields 10–15%. Although urban zones thus differ in albedo only slightly from their rural surroundings, the heat capacity of city structures is very high compared with soil and vegetation surfaces. This is because the stone and concrete of urban areas is very efficient at absorbing and storing heat received during the daytime. Later, when this heat is released (especially after sunset) to the air, it cushions the fall of nocturnal temperatures and therefore enhances the thermal contrast between urban and rural areas.

As a result of a more efficient drainage system and a generally impervious surface, urban landscapes are much drier than adjacent vegetation and soil surfaces. This means that much more of the available net radiation is used in heating the urban atmosphere directly than in evaporating moisture. When moisture is evaporated, latent energy is used which does not result in direct air temperature change. The warming effect caused by direct air heating and reduced evaporation over dry city surfaces is probably the main reason for the urban heat island effect.

(b)   Regional and global effects: cooling by desertification

In the dry lands of the world, drought, together with overgrazing, overcultivation and wood cutting, has been responsible for stripping large areas of land of their vegetation cover, and producing desert-like surfaces with higher albedos (Table 2.1 and Plate 2.1). One theory suggests that once desert landscapes are created (that is, in the process of desertification), they become self-perpetuating. This is because increased desert albedos result in a greater reflectance of incoming solar energy and a net loss of radiation at the surface. This produces a cooling effect and, since cool air is more dense than warm air and tends to sink, a general subsidence of air occurs. Such air subsidence limits the mechanisms associated with rising air currents, which are necessary for producing clouds and rainfall. Air subsidence therefore provokes a drying effect over the area.

It is feasible that humans may have altered surface albedos sufficiently to influence regional and global temperature patterns. Using a simple climatic model, Sagan *et al.* (1979) have calculated that global albedo changes during the last 25 years have been largely due to desertification. Moreover, such albedo alterations approximating to an increase of about 0.001% were capable of depressing global temperatures by 0.2°C. This calculation fits remarkably well with the actual decline in mean global temperature between 1940 and 1965 (see Figure 2.15). A continuation of present rates of land-use change and, in particular, of desertification suggests a further decline of 1.0 °C in the average global temperature by the end of next century. Sagan's results therefore suggest partial compensation for the increase in world temperature caused by the carbon dioxide greenhouse effect, anticipated from the continued burning of fossil fuels.

The impact of desertification on the energy balance of the planet may not be as straightforward as the foregoing analysis suggests. It should be kept in mind that the creation of desert surfaces is also associated with enhanced dust accumulations in the atmosphere. As outlined in the previous section, atmospheric dust may reinforce or help counteract cooling due to surface-related albedo change.

Figure 2.15 Changes in the average surface temperature (°C) of the atmosphere in the northern hemisphere (solid line) and southern hemisphere (broken line) between 1870 and 1975. (Source: Miller, 1979, p. E23)

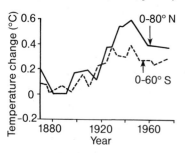

1. (a)   Describe the temperature patterns shown in Figure 2.14.
   (b)   Account for the temperature patterns in relation to: (i) nocturnal radiation cooling; (ii) atmospheric pollution emission; (iii) residual heat output; (iv) heat storage capacity; and (v) surface moisture modification.
   (c)   Discuss the relationship between the temperature patterns and (i) the size and shape of the urban area; (ii) the distribution of land and sea; and (iii) possible wind direction.

2. Refer to Figure 2.15 and the text.
   (a)   Describe the observed temperature variation of the northern and southern hemispheres between 1880 and 1975.
   (b)   Examine the possible role of global changes in (i) carbon dioxide levels, (ii) particulate output and (iii) surface albedos on the observed temperature trends in the northern hemisphere.
   (c)   What other factors might be involved in global temperature change?
   (d)   Indicate how temperatures are likely to alter in the future, if present trends in human activity continue.

Key Ideas

A.   *Energy in the Atmospheric System*

1. Solar radiation or insolation occurs as short-wave electromagnetic radiation and is the main source of energy input to the earth–atmosphere system.

2. The atmosphere is relatively 'transparent' to solar radiation, since it allows a large amount to pass through to the ground surface.

3. Nevertheless, only about one-half of the insolation available at the outer limit of the earth's atmosphere is actually absorbed by the earth's surface because of atmospheric reflection, scattering and absorption.

4. Insolation provides the energy responsible for all weather and climate, and for the growth of most living systems, and furnishes a vital supply of energy for the world's economy, indirectly through fossil fuels and hydro-power, and more directly from wind and solar power.

5. In contrast to solar radiation, terrestrial and atmospheric radiation are emitted at longer wavelengths and are known as long-wave or infra-red radiation.

6. The atmosphere is relatively opaque to infra-red radiation, since it absorbs a large proportion of the outgoing terrestrial radiation, sending it back to the ground surface as long-wave atmospheric radiation.

7. The raising of surface temperatures as a result of counter or back radiation from the atmosphere is known as the 'greenhouse effect'.

8. The net radiation balance at the surface of the earth is the difference between the absorbed solar radiation and the net outgoing long-wave terrestrial radiation.

9. The net radiation balance is employed in heating the ground, and subsequently the air above, by conduction, convection, turbulence and evaporation.

B. *Spatial Variations in Energy Characteristics*

1. There are marked spatial variations in the receipt of insolation at the earth's surface.
2. Latitude determines the intensity and thus the amount of annual insolation which, assuming a cloud-free atmosphere, shows a general decrease from the equator to the poles.
3. Because of the distribution of cloud cover, maximum solar radiation values are found in the cloud-free tropical and subtropical deserts.
4. For the earth–atmosphere as a whole, there is a surplus of net radiation in low latitudes (equatorwards of 40°) and a deficit in high latitudes (polewards of 40°).
5. As a result there is a transfer of heat from the lower to the higher latitudes.
6. There is also a gross transfer of heat from the ground to the air because the earth's surface, taken by itself, receives a positive (surplus) net radiation balance, whereas the atmosphere alone receives a negative (deficit) net radiation balance.
7. Despite latitudinal and vertical heat transfers, unequal patterns in the distribution of net radiation result in a general temperature decline from low to high latitudes, and from the ground surface through the atmosphere.
8. Unequal inputs and outputs of radiation (differences in heating and cooling) within the earth–atmosphere system are, in fact, the fundamental cause of all weather and climate.
9. A land surface tends to heat up and to cool down more quickly than an ocean surface.
10. This effect of 'continentality' can be explained in terms of:
    (a) the greater ease of penetration and distribution of the sun's rays and heat within the ocean;
    (b) the higher specific heat of water compared with land; and
    (c) the greater cooling effect of evaporation over the oceans.

C. *Temporal Variations in Energy Characteristics*

1. There are major temporal changes in the receipt of insolation and of net radiation at the earth's surface.
2. Large seasonal shifts in the distribution of solar and net radiation produce large latitudinal differences in hemispherical heating between January and July.
3. Latitudinal patterns in global heating are interrupted by regional variations introduced by the influence of land and sea distribution.
4. The average temperature difference between a station and its parallel, whether positive or negative, is its temperature anomaly.
5. Global temperature-anomaly maps, which emphasise regional vari-

ations in temperature, and thus the role of land and sea distribution, are shown to best effect during the seasonal extremes of January and July.

6. Using a simple model, daily temperature fluctuations in the air immediately above the ground surface can be related, under clear skies, to diurnal exchanges (gains and losses) of radiation.

7. The daily solar control of temperature at the ground/atmosphere interface is strongly modified by the introduction of cloudy skies, when winds blow from the sea on to a land surface, and by the presence of a snow cover.

8. Ground-based temperature inversions occur when there is an increase of temperature with height in the lower atmosphere, close to the ground.

9. At night, under clear skies and calm weather, an excessive loss of infra-red radiation from both ground surface and in the air immediately above, can result in a ground-based temperature inversion.

D. *Human Impact on Atmospheric Energy*

1. Humans can alter the energy balance and thus the temperature of the atmosphere over different space and time-scales by deliberate or unconscious action with beneficial or harmful results.

2. The energy balance is altered by:
   (a) changing the nature and composition of the atmosphere; and
   (b) altering the character of the earth's surface.

3. The urban heat island effect can be explained in part by atmospheric change through heat release and pollution emission from city areas.

4. The heat storage capacity and dry surface character of city structures also contribute to the urban heat island effect.

5. Recent global temperature trends show a general increase until around 1940, with an overall decrease or levelling off since then.

6. Increased anthropogenic accumulations of carbon dioxide have been implicated as a potential cause of global warming.

7. Such warming may enhance aridity in the mid-latitudes and cause rising sea levels due to glacier melt.

8. Particulate emission, whether by natural or anthropogenic means, may contribute to a global cooling effect.

9. Human modification of the earth's surface albedo, in particular by desertification, may also be responsible for planetary cooling.

Additional Activities

1. Examine the data shown in Table 2.5.
   (a) Using graph paper, plot values of insolation amount and sunshine duration at ground level (G, S) and at the top of the atmosphere (G_0, S_0) for Eskdalemuir and Trapani.
   (b) Compare and account for the monthly trends in insolation amount received at the top of the atmosphere (G_0) for the two

Table 2.5   Distribution of solar energy and sunshine duration at ground level and at the top of the atmosphere for selected stations

	Jan	Feb	Mar	Apr	May	Jun	Jul	Aug	Sep	Oct	Nov	Dec	Ann
Eskdalemuir (Scotland) 55° N 3° W, altitude 242 m													
G	4	11	20	32	39	46	41	34	23	13	7	3	23
G_0	17	32	56	83	104	114	110	91	66	40	21	13	63
S	9	24	31	45	48	59	47	44	32	24	18	12	33
S_0	78	96	118	141	162	173	168	149	127	104	83	72	123
Trapani (Sicily) 37° N 12° E, altitude 14 m													
G	24	33	46	60	75	80	84	76	58	42	29	21	53
G_0	46	61	80	99	111	116	113	103	87	67	50	42	81
S	46	52	62	75	93	100	113	104	83	71	57	42	75
S_0	98	107	118	130	140	145	143	134	123	111	100	95	120
London (England) 51° N 0° W, altitude 77 m													
G	5	11	21	30	41	50	44	36	27	16	8	5	25
S	13	25	38	47	63	75	63	59	49	34	23	14	42
Bracknell (England) 51° N 0° W, altitude 73 m													
G	6	12	23	33	44	52	47	38	29	17	9	5	27
S	12	24	37	47	60	73	62	57	48	32	23	13	41

G = mean solar radiation received on a horizontal surface at ground level (Wh/m^2 × 10^2)
G_0 = mean solar radiation received on a horizontal plane at the top of the atmosphere (Wh/m^2 × 10^2)
S = mean duration of sunshine falling on a horizontal surface at ground level (in tenths of an hour)
S_0 = mean duration of sunshine received on a horizontal plane at the top of the atmosphere (in tenths of an hour)
Source: Commission of the European Communities, 1979

stations in relation to (i) length of day (S_0), and (ii) the angle of the sun impinging on a horizontal surface.

(c) Explain the contrasting patterns of insolation received at ground level (G) at Eskdalemuir and Trapani.

(d) Compare solar radiation and sunshine values received at ground level in the city of London and in the nearby, but smaller town of Bracknell. Give reasons for your answers.

2. Refer to Figure 2.16 and the text.

(a) Describe the distribution of average January and July temperatures over the area shown.

(b) Given the general wind direction, and that clouds tend to form over warm surfaces rather than cold (due to rising air currents), where would you expect the greatest cloud cover to be in summer?

(c) Describe the spatial patterns of mean daily solar energy received at ground level in January and July.

(d) Explain these patterns with respect to latitude and the distribution of cloud cover.

(e) To what extent are seasonal temperature conditions determined by (i) the disposition of solar energy; (ii) the different reaction of land and sea to solar energy; and (iii) winds blowing from the sea.

Figure 2.16 Mean daily temperature (°C) and solar radiation received (kWh/m²) at the surface for January and July in the British Isles. (Source: Commission of the European Communities, 1979)

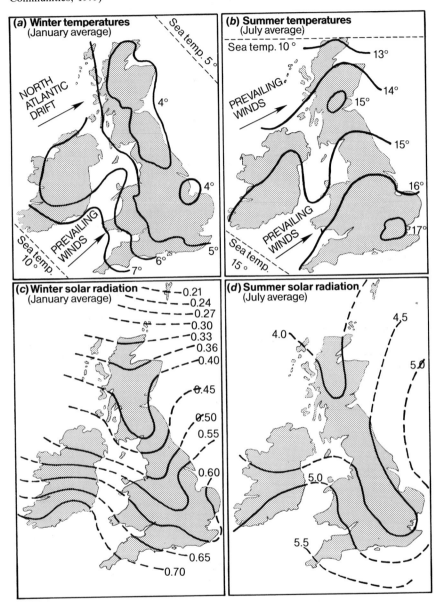

39

3 Atmospheric Motion

A. Importance of Air Movement

The atmospheric system is a dynamic system characterised by ceaseless air motion on a great variety of spatial and temporal scales. These movements range from minor short-lived events at scales of a few square centimetres or less to major seasonal motions encompassing areas of continental extent or greater. Accordingly, air movement is a vital determinant of weather and climate, and thus is of major human significance.

For convenience, air motion may be resolved into two components: horizontal and vertical. Horizontal movement, or wind, is by far the faster and consists of air movements parallel to the surface. Vertical motions, on the other hand, involve sinking and rising masses of air perpendicular to the surface and are usually 100–1000 times slower than their horizontal counterparts.

1. Horizontal movement

Horizontal movement is an important climatic factor to understand for a number of reasons. First, wind action physically relocates warm and cold bodies of air, thereby modifying the thermal characteristics bestowed upon places by their radiation regime. Such modification may have a considerable effect on the temperature of a place, as can be seen from Figure 3.1. The sudden drop in temperature which occurs for Chicago in midafternoon is associated with a change in wind direction, bringing cool air over Chicago and the lakeshore area from Lake Michigan. The temperature change brought about by this wind is clearly apparent for Chicago, unlike Joliet, which is situated further inland. Similar thermal effects can be observed over longer time-scales and over larger areas, illustrating the importance of wind as an agent of horizontal heat transfer, or *advection*. This was illustrated in Chapter 2 with reference to maritime influences.

Second, wind action transports water vapour, the source of life-giving precipitation. In particular, moisture is brought from areas where it is abundant, such as over the oceans, to areas where it is often deficient, such as over the interiors of continents. Figure 3.2 illustrates the significance of a seasonal reversal of wind direction in rainfall amounts for Bombay in India.

Figure 3.1 Course of temperature at Chicago and Joliet, Illinois, during one 24-hour summer period. (Source: Oliver, 1979, by permission of J. E. Oliver)

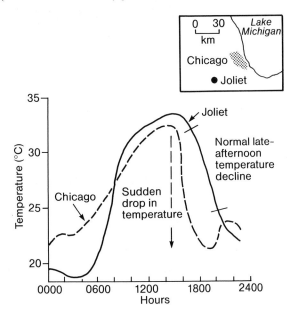

Figure 3.2 Monthly rainfall and wind direction at Bombay

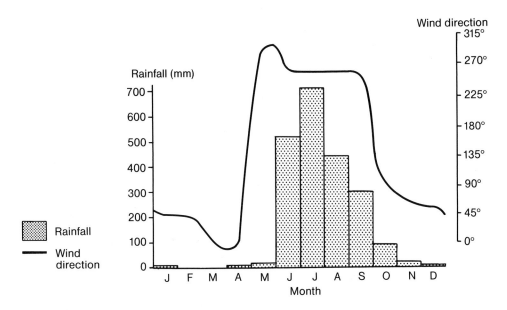

Third, air in rapid motion is, on occasion, a severe environmental hazard. On average, more lives are lost each year as a result of tropical storms than from the combined effects of fire, lightning, floods, tidal waves and earthquakes.

2. Vertical movement

Vertical movements of air, although normally less rapid than their horizontal counterparts, are no less important, since they strongly influence whether the climate and weather will be cloudy and rainy or clear and dry. For reasons that will be dealt with in Chapter 4, areas where air is predominantly sinking are relatively cloud-free and dry, whereas in areas characterised by rising air motion the opposite weather types tend to prevail.

3. Air motion and the global energy budget

Air in motion, however, has an even more fundamental function to fulfil at a global scale. It will be recalled from section B of Chapter 2 that the unequal heating of the earth's surface by the sun produces a latitudinal contrast in energy budgets. Between about 40° N and 40° S, the amount of incoming radiation exceeds that lost by the cooling of the earth–atmosphere system, whereas towards the poles the reverse applies. Obviously, if such a situation persisted over very long time periods, it would cause the low latitudes to be very much hotter than they are at present, and the high latitudes to be very much more frigid. The fact that this is not so implies the existence of a mechanism whereby heat is moved from the surplus areas to the deficit areas to compensate for the shortfall in the energy budget of the latter. This vital role, without which only a small proportion of the globe would be habitable, is fulfilled by a global system of wind circulation, described later in this chapter.

ASSIGNMENTS
1. (a) *What is meant by the term 'advection'?*
   (b) *With reference to Figure 3.1, suggest wind directions which may have been experienced at Chicago at:*
       (i) 0600   (ii) 1200   (iii) 1800
   (c) *From what compass directions are the following winds blowing:*
       (i) 90° (ii) 180° (iii) 225° (iv) 025°?
2. (a) *In what other ways does wind action influence climatic conditions at a place?*
   (b) *Make a list of important non-climatic functions fulfilled by the wind.*

B. Forces that Control Atmospheric Motion

Atmospheric motion is controlled by the interplay between four forces: (i) the pressure-gradient force; (ii) gravity; (iii) the Coriolis force; and (iv) friction.

1. Pressure-gradient force

Motion is a response to a force or forces of some kind. This is one of Sir Isaac Newton's two basic laws of motion, which state, first:

An object (such as a parcel of air) at rest remains at rest unless acted on by an unbalanced force, and an object in motion continues its motion in a straight line at constant speed unless acted on by an unbalanced force.

And second:

The acceleration of an object of unit mass is directly proportional to the sum of forces acting on it.

(a) Air pressure

In moving about randomly, gas molecules collide with each other and with any surface which confines them. The push exerted by this continuous bombardment is known as *pressure*. The atmosphere is a mixture of gases, confined by the ground or sea below and the force of gravity, which prevents escape above. Although not readily noticeable, air exerts a pressure on every surface exposed to it. That pressure can be considered as resulting from the weight of overlying air pressing down on a given area. Obviously, the layers closest to the surface will have the greatest weight overlying them and thus the pressure will be greatest, and vice versa for the layers at the top of the atmosphere. Table 3.1 shows the rapid decrease in air pressure which occurs with increasing height. For each 5 km rise in altitude, pressure decreases by about a half; for example, at a height of 5 km, it is one-half of its sea-level value, at 10 km a quarter, at 15 km an eighth, etc.

Table 3.1   Relationship between air pressure, temperature and altitude

Height (km)	Temperature (°C)	Pressure (mb)	Height (km)	Temperature (°C)	Pressure (mb)
0	15	1013.2	9.0	−43	308.0
0.5	12	954.6	10.0	−50	265.0
1.0	9	898.8	12.0	−57	194.0
1.5	5	845.6	14.0	−57	141.7
2.0	2	795.0	16.0	−57	103.5
2.5	−1	746.9	18.0	−57	76.65
3.0	−4	701.2	20.0	−57	55.29
4.0	−11	616.6	30.0	−47	11.97
5.0	−17	540.4	40.0	−23	2.87
6.0	−24	472.2	50.0	−3	0.78
7.0	−30	411.1	60.0	−17	0.23
8.0	−37	356.5	70.0	−53	0.06

(b) Measurement of air pressure

The unit of force most widely used in physics is the newton (N): 1 N is the force that would accelerate 1 kg of mass by 1 m/s. Measuring air pressure in these units, it is found that, on average, the force exerted at sea level is about 101 325 N/m^2. Fortunately, air finds its way into so many locations – within liquids, porous surfaces, even our own bodies – that this force is often exactly counterbalanced and we are not constantly conscious of its existence.

Air pressure is seldom expressed in terms of newtons per square metre (N/m^2); meteorologists prefer to use a unit known as the *millibar* (mb), where 1 mb = 100 N/m^2. Accordingly, the standard average sea-level pressure shown in Table 3.1 is 1013.2 mb.

Historically, air pressure has been measured in units other than millibars. This stems from a classic experiment performed by Torricelli in 1643. Torricelli filled a narrow tube with mercury, and inverted it in a dish of mercury. He reasoned that the length of the mercury column in the tube, about 76 cm, was a measure of the pressure of air pressing down on the mercury in the dish. Here then was a device for measuring air pressure in terms of centimetres or inches of mercury. Although today's mercury barometers are rather more sophisticated, they embody the same principles as Torricelli's original device.

Other instruments for measuring air pressure have also been developed. In particular, the use of a sealed container, with slightly flexible walls from which most of the air has been removed before sealing, provides a device sensitive to changes in air pressure. This is called an *aneroid barometer*, and it can be connected to a rotating drum and be used to provide a daily or weekly record of changes in air pressure. The range of air-pressure variation at the surface seldom exceeds 50 mb above or below the 1013 mb average. The highest air pressure recorded was 1084 mb, observed at the centre of a large area of high pressure that was located one winter over the frozen heart of Siberia. The lowest value ever recorded at the surface was 876 mb, close to the eye of a tropical storm in the Pacific Ocean.

(c) Mapping air pressure

Air-pressure values measured simultaneously at a number of locations may be mapped. Because these places are all situated at different heights above sea level, the values are adjusted to give the equivalent sea-level pressures. This enables us to see, for a common vertical height, where differences in pressure exist between places. As an aid to this, it is useful to draw in lines which connect the places that share the same value of air pressure. Such isolines are known as *isobars*. Figure 3.3 shows an isobaric map from which the differences in air pressure between places can be easily interpreted.

Where the isobars curve round to totally enclose an area of low pressure, this is called a *low* or *depression*. Where the isobars enclose an area of high pressure, the opposite terms *high* or *anticyclone* are used. Elongated areas of high or low pressure, without total enclosure by the isobars shown, are described as *ridges* or *troughs* respectively.

Isobars enable the places where pressure changes rapidly over short distances to be identified. The closer together the isobars are, the greater the pressure gradient. Just as tightly packed contours indicate a steep topographical gradient, so tightly packed isobars are indicative of a steep air-pressure gradient between places. Line AB in Figure 3.3 shows a distance over which there is a relatively large pressure gradient, whereas line CD shows a similar distance over which a gentle pressure gradient prevails.

Figure 3.3 Surface isobaric chart (in mb) for 0600, 5 December 1980

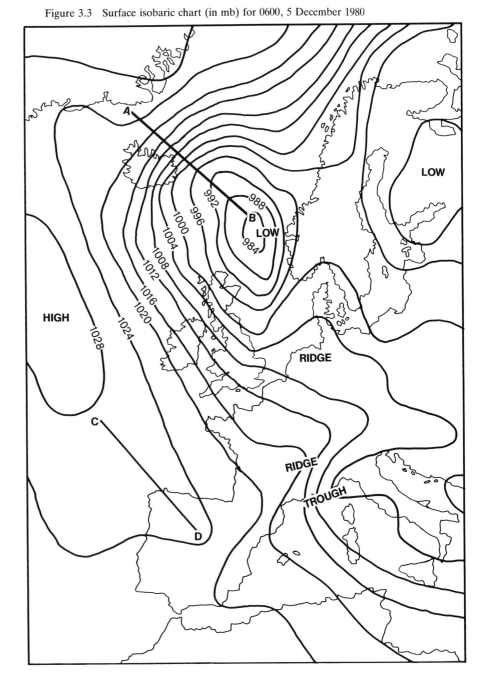

The analogy with surface relief may be carried further, because a steep pressure gradient, like a steep slope, causes more rapid acceleration of material down the gradient than does a more gentle gradient. A simple relationship between pressure gradients and wind speed thus exists: the steeper the pressure gradient, the faster the wind speed.

Isobars may also be drawn to show variations in pressure at given heights above the surface, apart from the sea-level case considered above.

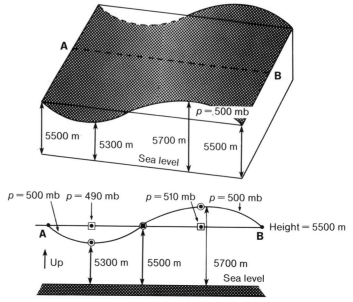

Figure 3.4 Depiction of the changing height of an imaginary 500 mb surface. (Source: Trewartha and Horn, 1980)

For these exercises, the pressure distribution may be depicted slightly differently, to show at what particular height a given pressure value is reached. For example, the pressure value of 500 mb may be located 5300 m above one point and 5700 m above another. Pressure at the surface will usually be higher for the latter location, since the decline in pressure with height requires a greater distance to reach the 500 mb level. It is possible thus to visualise an imaginary 500 mb (or 700 mb, or 850 mb) isobaric surface, gently rising and falling over an area, which joins up all the points at which the pressure equals 500 mb. Figure 3.4 shows a representation of the changing height of the imaginary surface. These heights could be displayed as contours, as in Figure 3.19. Such maps provide information on pressure gradients at high levels in the atmosphere, which is often vital in understanding and predicting weather and climatic events several thousands of metres below at the surface.

2.   Force of gravity

It can be concluded from the previous section that the greatest pressure gradients occur vertically in the atmosphere, not horizontally. This would appear to suggest that the atmosphere should shoot away from the earth into the vacuum of space. The fact that this does not happen is because the vertical pressure-gradient force is almost always balanced by the *force of gravity*, acting in the opposite direction. These two great forces cancel each other out most of the time to produce a state of stability known as *hydrostatic equilibrium*. Relatively small imbalances occur between them, producing rising and sinking motions; these motions are usually much more gentle than their horizontal counterparts, where no comparable check to the pressure-gradient force exists.

3. Coriolis force

(a) *The nature of the Coriolis force*

A satellite, stationary over the North Pole, would have a view of the earth not unlike that shown in Figure 3.5(*b*). From its fixed position, its sensors would see the earth below complete one revolution in an anticlockwise direction every 24 hours. Now imagine a rocket is fired from the surface at the North Pole towards some target near the equator (*X*). If the rocket takes 2 hours to reach its destination, during the time it was airborne the earth would have rotated 30° to the east. Accordingly, the rocket would

Figure 3.5   The Coriolis force. (Source: Lutgens and Tarbuck, 1979). The curved arrow shows the path taken by the rocket with respect to the earth's surface

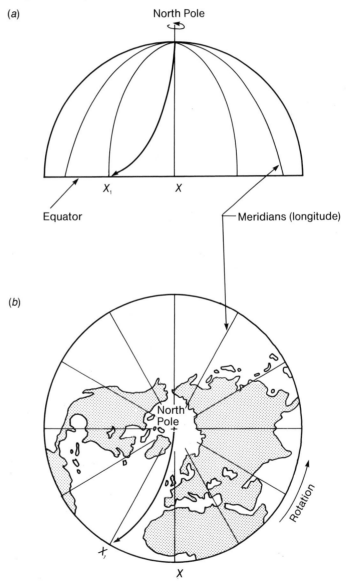

land at X_1 some 30° to the west of its intended target. To anyone viewing the operation from the earth, the trajectory of the rocket would have traced out a curved path towards the west. From the satellite, however, a different conclusion would be drawn: namely, that the rocket flew in a straight line, but the rotation of the earth resulted in it missing its target.

Since we measure motion from our position on the earth's surface, rather than from some fixed point in space, this complication due to the spinning earth must be allowed for when studying objects which move with respect to the surface. The deflection which they apparently experience is termed the *Coriolis deflection* and can be considered as a consequence of a force acting to the right of the direction of motion in the northern hemisphere and to the left in the southern hemisphere.

(b) Attributes of the Coriolis force

Further consideration of Figure 3.5 will suggest four conclusions regarding the Coriolis force.

1. The Coriolis force acts at right angles to the direction of a moving object, towards the right in the northern hemisphere and towards the left in the southern hemisphere.
2. The Coriolis force influences only the *direction* of motion, not the speed.
3. At a given latitude, the Coriolis force increases as the speed of the object increases. (An object which appears stationary or very slow-moving to us on the earth is obviously moving in unison with the rotation of the earth and experiencing no, or very little, deflection.)
4. The Coriolis force is strongest at the poles and zero at the equator. (The 'twist' of the turning earth is greatest at the equator.)

4. Force of friction

Very high wind speeds are most often observed over the sea. This is because air moving over land is subjected to a greater frictional drag, which slows it down, particularly if the surface is rough and irregular. Large forests or tall buildings slow down the wind to a much greater extent than does a relatively smooth sea surface. All types of obstacles protruding into the air contribute to this frictional drag, which is obviously greatest close to the surface. Above a height of about 1000 m this slowing of wind becomes slight – except on hot afternoons, when rising towers of air from the heated ground below obstruct the motion of the air aloft, rather like a skyscraper skyline would. Friction has important consequences not just for wind speed, but also for wind direction, as will be apparent later.

ASSIGNMENTS
1. *(a) What force is responsible for generating wind motion?*
   (b) Define the terms 'millibar', 'isobar' and 'isobaric surface'.
   (c) What is the standard air pressure at sea level:
      (i) in millibars; (ii) in centimetres of mercury?

(d) *On an isobaric chart describe how you would distinguish (i) a depression from an anticyclone; (ii) a ridge from a trough.*

(e) *What relationship does wind speed at the surface usually show with the spacing of isobars on a chart of sea-level air pressure?*

(f) *Why do vertical pressure differences not generate great vertical motions in the atmosphere?*

2. (a) *Explain how the Coriolis force results in an apparent deflection of moving objects.*

(b) *How does latitude and the speed of the object influence the magnitude of the Coriolis force?*

(c) *In the northern hemisphere, if you are facing away from the wind, is the low-pressure area on your right or left-hand side?*

C. Scales of Atmospheric Motion

1. Microscale movement

(a) *Winds above the friction layer*

Above the influence of surface frictional drag, air movement is controlled by two of the forces already described: the horizontal pressure-gradient force and the Coriolis force. For the purpose of explaining horizontal air movement, or wind, the forces of gravity and the vertical pressure gradient may be assumed to cancel each other out and may be ignored.

Figure 3.6 shows how a parcel of air in the northern hemisphere would respond to the two controlling forces. From its starting point, the air would begin to move in response to the pressure gradient from high to low pressure. Once it begins to move, however, it becomes subject to the Coriolis force, which displaces it to the right of its trajectory. As the parcel speeds up as a result of the continued presence of the pressure-gradient force, the Coriolis force also intensifies. (The Coriolis force increases as the speed of an object increases.) As the parcel of air accelerates, the magnitude of the deflection grows, until the air parcel is moving parallel to the isobars. At this stage the two forces are acting in

Figure 3.6   Forces controlling wind direction aloft

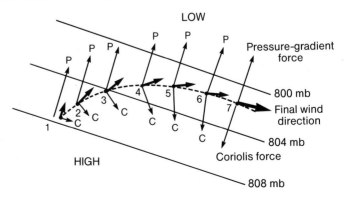

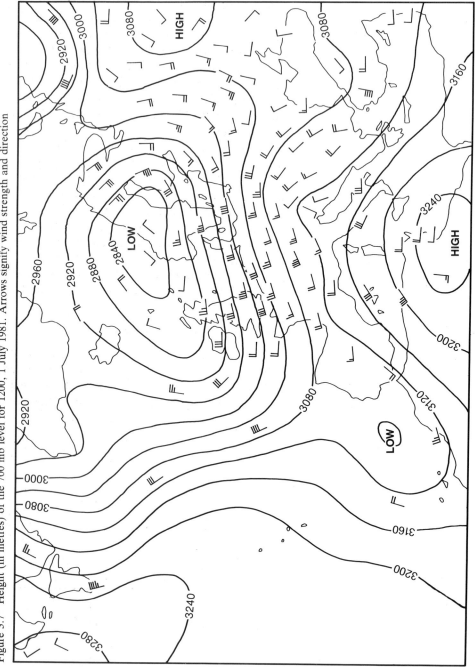

Figure 3.7 Height (in metres) of the 700 mb level for 1200, 1 July 1981. Arrows signify wind strength and direction

opposite directions and, since neither is gaining the upper hand in dragging the air parcel in its direction, the two forces must be balanced. The result is that the air parcel continues to move parallel to the isobars.

Since no unbalanced force is now acting on the air parcel, it continues to move at a constant speed and in the direction it was travelling when balance was achieved. The speed reached will obviously depend on two things. First, how strong the initial pressure gradient was. The steeper the initial pressure gradient, the faster the air parcel would have been travelling before balance was achieved. Second, how strong the Coriolis force was. If the Coriolis force was relatively weak, as in low latitudes, the parcel of air would again have reached a higher speed before a balance was achieved.

The wind produced as a result of this balance between the pressure-gradient force and the Coriolis force is known as the *geostrophic wind*. Such winds, blowing parallel to the isobars, are a feature of the upper air circulation. Figure 3.7 shows a chart of the isobars at the 700 mb level for part of the northern hemisphere. The wind direction arrows can be seen to be running parallel to the isobars. In addition, they indicate that the wind blows in a clockwise manner around areas of high pressure, and in an anticlockwise manner around areas of low pressure. This is an important relationship, which results from the interaction of the two forces described in this section; it will be considered in greater detail below.

(b) Winds within the friction layer

One consequence of the frictional drag is that it disrupts the balance between the horizontal pressure-gradient force and the Coriolis force, described above for the higher layers of the atmosphere. In so doing, this friction in the lower layers alters wind direction. By reducing the speed

Figure 3.8 Forces controlling wind direction at the surface

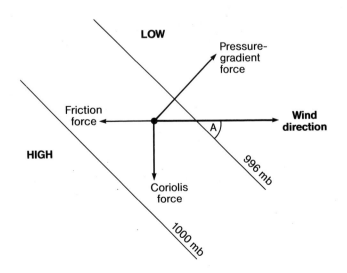

of moving air, friction also reduces the magnitude of the Coriolis force which, as has already been demonstrated, is dependent on wind speed. Since the pressure-gradient force is unaffected by wind-speed changes, the contest between it and the Coriolis force becomes unequal, and the pressure-gradient force becomes the dominant partner. In response to this now unbalanced force, winds blow obliquely across the isobars in the direction of low pressure (Figure 3.8).

The angle at which the wind will cross the isobars will vary according to how much frictional reduction of speed takes place. Over an ocean surface, for example, only a slight reduction of the Coriolis force will occur, and the air will flow across the isobars at a relatively small angle, 10–20°. Over very rough surfaces, where the wind speed close to the surface may be halved by friction, the angle will be greater: 45° or more.

The effects of friction in changing wind direction are especially important for two types of pressure systems: depressions and anticyclones. In a depression, the crossing of the isobars by winds coming from a number of different directions produces a net inflow of air at the centre. This may only escape upwards (see Figure 7.6). Conversely, a net outflow of air at the centre of an anticyclone necessitates air being drawn down from aloft at the centre. Thus an important relationship is apparent between low-pressure (or cyclonic) flow, *convergence* at the surface, and upward motion at the centre of a depression. Equally, high-pressure (or anticyclonic) flow is associated with *divergence* of air at the surface, and *subsidence* of air from higher up at the centre of the anticyclone. These relationships will be returned to later.

2. Mesoscale motion

Air-pressure differences cause motion on a variety of scales (Table 3.2). In many ways atmospheric motion may be likened to a fast-flowing river: swirls and counter-currents at small scales may be seen to be part of bigger eddies at larger scales, and so on. Each scale is closely connected with the other ones in what is ultimately a unitary system driven by some form of energy input. Although scales of analysis are arbitrary, atmospheric motions that extend for horizontal distances of 1–100 km are often termed *mesoscale*. A good example of this type of motion is provided by the land/sea-breeze circulation.

Table 3.2 Temporal and spatial scales for atmospheric motions

Name of scale	Time-scale	Length scale	Example
Macroscale			
General circulation	Weeks–years	1000–40 000 km	Waves in westerlies
Synoptic scale	Days–weeks	100–5000 km	Cyclones, anticyclones, hurricanes
Mesoscale	Minutes–days	1–100 km	Land–sea breeze, thunderstorms, tornadoes
Microscale	Seconds–minutes	<1 km	Turbulence

Source: Lutgens and Tarbuck, 1979

(a) Land and sea-breeze model: thermal aspects

It will be recalled from Chapter 2 that land and water surfaces differ markedly in their temperature response to solar radiation. The land surface, warming quickly early in the day, warms the columns of air overlying it. The sea surface, by contrast, warms much more sluggishly, producing a difference in temperature at similar heights in the air above both surfaces. Warming of the air column over the land causes its expansion and increases the vertical distance between isobaric surfaces. For example, the vertical distance between the 1000 mb and 950 mb isobaric surfaces would be enlarged because of the expansion of the land-based air column. The result of this is that, at similar levels aloft, air pressure over the sea is slightly lower than over the land, and air begins to drift seawards aloft in response to this. The removal of air from the land-based column and its addition to the sea-based column produces a reduction in air pressure at the surface over the land and an increase over the sea surface. At the

Figure 3.9   Formation of a sea breeze. (Source: Oliver, 1979, by permission of J. E. Oliver)

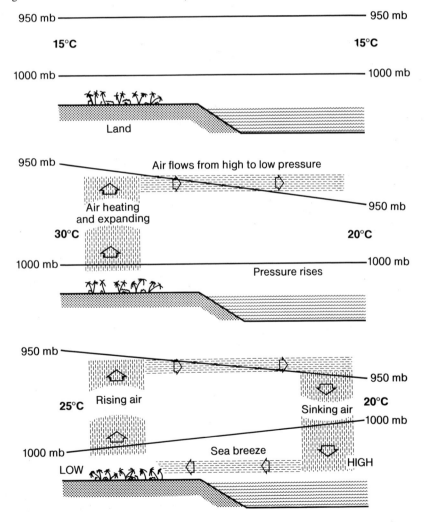

surface, therefore, a pressure gradient now exists from the sea to the land, and it is this which results in the onset of the onshore breeze so characteristic of coastal areas on warm summer days. A convective cycle of air motion has been established with air moving onshore at low elevations, being warmed and induced to rise, flowing offshore aloft where it cools and sinks to begin the cycle again (Figure 3.9). Removal of the temperature differential between the land and sea surfaces (for example, as the land cools rapidly in the evening) will result in the cessation of the sea breeze. Reversal of the temperature gradient, as occurs during the night when the sea is warmer than the land, may produce an offshore flow at the surface, known as the land breeze.

(b)   Land and sea-breeze model: moisture aspects

The distance to which a sea-breeze circulation extends inland is often identifiable by a line of convective clouds, where the upward motion of the moist oceanic air leads to condensation and cloud formation. This is very clearly displayed on the satellite photograph of part of the coastline of East Africa (Plate 3.1). The coastal margin of Mozambique and Tanzania is distinctly cloud-free, whereas further inland extensive cumulus development is apparent. The edge of this cloud runs parallel to the coastline, showing the extent to which the sea breeze has progressed inland. Such a boundary has sometimes been termed a *sea-breeze front*. Sea breezes also occur around large lakes, and these are also apparent on the satellite photograph. On occasion, sea breezes from more than one direction appear to converge on each other, enhancing the convection and leading to more vigorous cloud formation.

(c)   Valley winds

Different rates of heating and cooling are also responsible for a mesoscale circulation occasionally observed in valley settings. During the night, colder, denser air at higher elevations drains gently downslope under gravity at speeds of about 1 m/s. Upon reaching the valley floor, a movement towards the lower ground along the valley axis occurs. These flows of cold air are very shallow and easily obstructed by obstacles such as forests, walls, buildings, etc., behind which ponding may develop before overflowing occurs. Figure 3.10 shows the arrival of episodic bursts of cold air at lower elevations in a valley in central Scotland during one night in 1976. Such downslope movements of air are known as *katabatic winds* and may be marked over surfaces such as glaciers, where intense chilling of air occurs. In relatively enclosed low-lying areas, katabatically induced chilling of the air overlying the surface renders such areas more susceptible to fog and frost. Farmers therefore should try to avoid valley-floor locations for frost-sensitive crops and plant them just above the level at which the katabatic pond (frost pocket) forms.

During the day, a reversal of this circulation occurs. Air moves upslope and up along the valley axis, in response to the greater heating of the air at lower elevations. This latter circulation is known as an *anabatic wind*.

Plate 3.1 Meteosat image of part of East Africa. Sea breezes are prevalent along many tropical coasts. Here, the warm African land mass has been instrumental in creating extensive sea-breeze activity. Note how the line of cumulus clouds marking the inland limit of moist maritime air parallels the coastline, and how the large inland lakes are also responsible for the sea-breeze effect on a smaller scale. (METEOSAT image supplied by the European Space Agency)

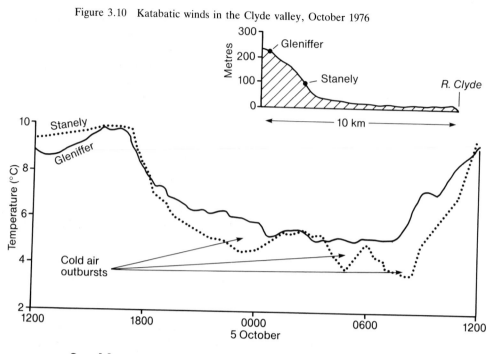

Figure 3.10 Katabatic winds in the Clyde valley, October 1976

3. Macroscale motion

(a) The global circulation model: thermal aspects

Heating and pressure differences have been shown to be responsible for the convective cycling of air on coastal margins, which we termed the sea-breeze circulation. However, such thermally induced pressure differences also exist at much larger scales. Spatial variations in the energy budgets of places were examined and Figure 2.6 clearly implied the existence of a fundamental contrast in heating at a global scale. In terms of the relationship between radiation energy received and lost, two very different categories could be identified. Equatorwards of 40° N and 40° S, a zone of surplus energy existed for the earth–atmosphere system as a whole, whereas polewards of these latitudes a zone of energy deficit was apparent, one in each hemisphere. This imbalance comes about mainly because of variations in incoming energy intensity, related to the angle of orientation of the earth's surface to the sun and to the characteristics of its orbit around the sun. These features are not unique to the earth; nor, therefore, is the radiation imbalance between high and low latitudes, which they cause. For example, a similar situation probably exists on the planet Venus.

A place losing more energy than it receives in the course of a year would suffer a fall in its mean annual temperature, and vice versa. Over a long period of time, the contrast in temperature between high and low latitudes on the earth would, one might expect, become more and more pronounced. Perhaps the only habitable place on the earth would be a narrow zone in the middle latitudes, sandwiched between a frigid wasteland and parched tropics: unless, that is, some mechanism existed whereby the excess heat of the low latitudes could be transported towards the poles to subsidise the heat-deficient areas of the higher latitudes. Fortunately, such a mechanism does exist in the motion of the winds and ocean currents. These are activated on a global scale by the latitudinal heat imbalance and the pressure contrasts to which it gives rise. A planetary-scale convective circulation is the atmospheric response to the spatial inequality in energy budget analysed in Chapter 2.

(b) The global circulation model: pressure aspects

Average sea-level air pressures for January and July are shown in Figures 3.11(a) and 3.11(b). At both these times, prominent pressure features are apparent at particular latitudes and they may thus be described as semi-permanent. Foremost among these is the belt of lower-than-average press-

Figure 3.11   Global surface air pressure (in mb) (a) January; (b) July

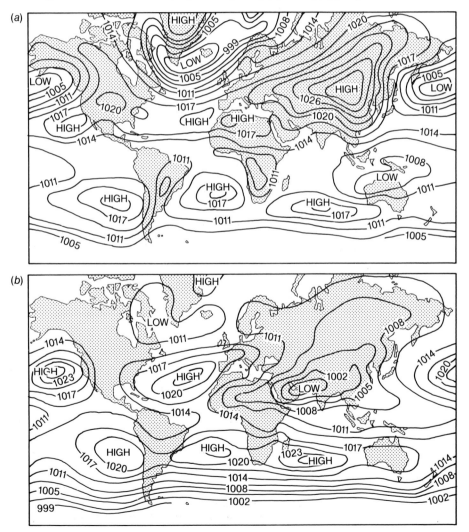

ure in the vicinity of the equator. This is the *equatorial trough*, induced by the rising air motions which characterise that intensely heated part of the globe. In contrast, at about 30° N and 30° S lie the *subtropical high-pressure cells*, especially well developed over the oceans in summer. These anticyclones are associated with descending air motions and give way to lower-pressure zones on their poleward margins.

A seasonal migration polewards in pressure belts is also apparent. In the northern hemisphere, in particular, this is complicated by the juxtaposition of oceanic and continental areas, and results in high-pressure centres developing in winter over large land masses. This is most apparent over Eurasia, and contrasts sharply with the low-pressure systems over the warm mid-latitude oceans – for example, the Icelandic and Aleutian Lows. In summer, on the other hand, the warmer land masses generate low pressure, while enhancement and poleward expansion of the subtropical highs – for example, the Azores and Hawaiian Highs – occurs at the expense of the aforementioned low-pressure areas.

(c)   *The global circulation model: ocean currents*

The semi-permanent pressure features described above are responsible for the prevailing wind regimes experienced over the globe. These in turn drive surface ocean currents, which are also important thermal regulators of the global energy budget, exchanging heat between the low and high latitudes and, in so doing, influencing strongly the climatic characteristics of places. Figure 3.12 shows the surface drifts and currents of the oceans in January. Clearly the great circular movements, or *gyres*, are controlled by the subtropical highs shown in Figure 3.11.

Figure 3.12   Generalised circulation of the oceans in January

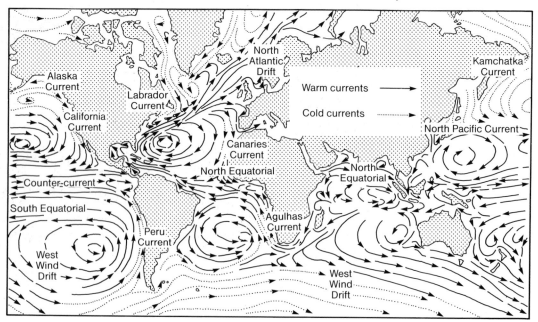

ASSIGNMENTS

1. (a) *Explain how and where geostrophic winds occur.*
   (b) *Describe how frictional forces induce convergence of air at the surface in a depression, and divergence at the surface in an anticyclone.*
2. (a) *Graph the following observations of temperature and relative humidity (RH) which were made at a coastal location in Devon, England, in early summer. (See Chapter 4, p. 76, for a definition of relative humidity.)*

Time:	0900	0930	1000	1030	1100	1130	1200	1230
°C:	11.0	11.4	11.9	12.6	13.3	14.3	16.5	13.2
RH%:	60	58	51	46	42	39	41	59

Time:	1300	1330	1400	1430	1500	1530	1600	1630
°C:	13.3	13.5	13.8	13.8	13.5	13.0	12.8	12.2
RH%:	59	58	55	55	58	58	57	58

   (b) *Explain in detail the reasons that may underlie the abrupt changes in temperature and humidity that are apparent on the above graph.*
3. *Plate 3.2 shows a satellite image of part of the Mediterranean basin.*

Plate 3.2 Meteosat image of the Mediterranean basin. (METEOSAT image supplied by the European Space Agency)

(a) *Describe and explain the linear cloud formations close to the arrowed areas on coastal margins.*

(b) *Suggest why much more active and clustered cloud formation appears to be occurring over Greece.*

D. Modelling the General Circulation of the Atmosphere

1. Early circulation models

The concept of a thermal circulation powered by the temperature contrast between the poles and the equator can be traced back at least 300 years. Since then, the rather simple early models of the circulation have been superseded by increasingly sophisticated versions which correct earlier weaknesses or misconceptions. Such flaws have been highlighted by the modern existence of a greatly expanded observational network, and by new sources of data, especially for motion aloft. Even so, it must be admitted that our understanding of the workings of the global circulation is still far from complete.

(a) Motion on a stationary earth: Halley's model

In 1686 Edmund Halley proposed the existence of two great convective cells, one in either hemisphere, driven by rising masses of air overlying the area of most intense solar heating close to the equator. As these masses of warm air rose, they would cool by radiation to space and lose their buoyancy. Encouraged by the continuing updraught from below, they would begin to flow towards the poles, undergoing further cooling as they went. Ultimately, their increased density would induce them to sink back to the surface, creating a high-pressure area from which a return flow at the surface back to the equatorial low-pressure zone would complete the convective cell (Figure 3.13).

Figure 3.13   Unicell circulation model – Halley: a thermally direct cell on a stationary earth

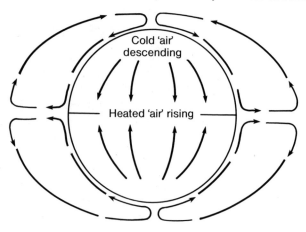

(b) Motion on a rotating earth: Hadley's model

Halley's scheme was modified in 1735 by George Hadley,
include the effects that the earth's rotation would have on ~
He reasoned that the poleward-flowing currents of air aloft would be de
flected to the right in the northern hemisphere and to the left in the southern hemisphere to become south-westerlies and north-westerlies
respectively. More importantly, the surface return flows would become
north-easterlies and south-easterlies in the northern and southern hemispheres respectively (Figure 3.14). Formulated during the era of sailing
ships, such a model seemed to explain well the reliable winds from these
two directions which mariners consistently encountered in the tropics and
which they had dubbed the 'trade winds'. Although Hadley's model was
later challenged, the direct circulation principle which it embodies, or the
'Hadley cell', remains an important component of modern circulation
models, particularly for the circulation features of the low latitudes.

One major flaw exists in Hadley's model. In both hemispheres, the
surface wind blows towards the west, opposing the direction of the earth's
rotation. The wind regime is thus capable of exerting a frictional drag on
the motion of the earth. If there was in the model an equal eastward-
flowing component, this frictional force would be cancelled out. But there
is not. Therefore acceptance of Hadley's model implies acceptance of an
unbalanced force which, over time, would have the effect of slowing down
the rotation of the earth. This conclusion follows from the first of Newton's laws, (see section B above). Fortunately, there is no evidence of
such a slowing down occurring, and so our search resumes for a new
model that will incorporate a better balance between eastward and
westward-blowing surface winds.

(c) Tricellular models: the models of Ferrel and Rossby

A suggestion involving three cells in each hemisphere was made in 1856
by Ferrel and elaborated on by Rossby in 1941. In this scheme, two Hadley cells were advocated (see Figure 3.15).

(i) *Low-latitude direct cell.* This extends from the equator to about 30°
latitude and is driven by the warm, rising masses of air in the equatorial
zone. The rising and consequent cooling of these warm, normally humid,
air masses induces them to shed their excess water vapour as cloud. This
releases large amounts of heat energy, the latent heat of evaporation,
which increases the instability of the air masses, producing further uplift
and cooling. The large-scale formation of cumulus cloud is responsible in
turn for the copious rainfall in these areas (see Plates 2.1 and 2.2).

On moving towards the poles aloft, this air encounters a progressively
stronger Coriolis force, until by 25° latitude the current is deflected almost
in a west-to-east direction. Further poleward movement is rather restricted therefore, and a piling-up of air aloft occurs at these latitudes.
This convergence of air aloft, allied to its cooling and loss of buoyancy,
leads to subsidence in the zone between 20° and 35° latitude. As downward motion occurs, the air warms under the compressive effect of the

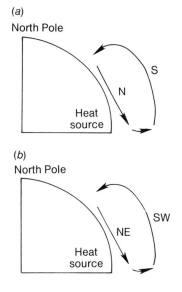

Figure 3.14 Unicell circulation model. (a) Halley: a thermally direct cell on a stationary earth. (b) Hadley: a thermally direct cell on a rotating earth

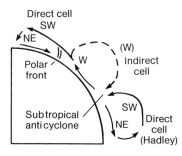

Figure 3.15 The tricellular circulation model

61

pressure exerted by the air above. These areas of subsidence, the subtropical anticyclones, correspond to the desert regions of the world. Over the oceans, the light winds in this zone of predominantly descending air becalmed the mariners of old, forcing them to jettison horses and other livestock on board for which they could no longer provide water; hence the name 'horse latitudes' was often used to describe these regions. Between the 'horse latitudes' and the equator, a return flow of air at the surface exists, deflected by the Coriolis force to give the north-easterly trade winds in the northern hemisphere and the south-easterly trades in the southern hemisphere.

(ii) *High-latitude direct cell.* A second, and much more restricted, direct cell was postulated as existing in high latitudes, driven by the chilling-induced subsidence of air immediately over the polar ice-caps. Surface outflow from this small polar anticyclone was restricted, however, by the presence of a further cell, different in nature from the direct cells already described.

(iii) *Mid-latitude indirect cell.* Between the horse latitudes and the rising limb of the high-latitude cell, Rossby suggested the existence of a third cell, driven by friction with the two direct cells adjacent to it. On the equatorward side, this friction would be provided by the descending motion of the air within the subtropical highs. Some of this air would also 'spill' polewards to produce (because of deflection) the westerlies of the middle latitudes. On the poleward side, the upward limb of the polar cell could be envisaged as providing further frictional energy, enabling a return flow aloft equatorwards to occur within this 'indirect cell'. The poleward boundary of this indirect cell marked an important junction between tropically-derived air and the cold polar circulation. It was later included in the model as the 'polar front'.

(d) *Tricellular model: a reappraisal*

The tricellular model depicted in Figure 3.15 represented a considerable advance on the earlier generations, particularly in its ability to explain the surface wind directions in the middle latitudes. However, within eight years of presenting his scheme, Rossby was forced to abandon most of it. Like its predecessors, it too had a fatal flaw.

Consideration of Figure 3.15 will suggest that the return flow aloft for the indirect cell should be a north-easterly flow, due to Coriolis deflection of an equatorward-moving current. During the 1940s, however, more sophisticated means of investigating the upper atmosphere, particularly the use of the radiosonde, were more extensively deployed. These demonstrated that, instead of reversing aloft, the mid-latitude westerlies increased in persistence and strength, reaching velocities of up to 400 km/h in extreme cases. In addition, the measured amount of heat transferred by convective overturning between low and high latitudes was found to be only a fraction of what must take place to balance the earth's energy budget. Clearly, a fundamental weakness afflicts all the generations of models considered so far: it lies in the premise on which they were constructed. All assumed that the earth's energy budget was balanced by convective motions, that is, overturning of air in great global cells. Could it be that another form of heat redistribution, controlled from the mid-latitude area, is also an important component?

2. Modern circulation models

(a) *Dishpan experiments*

New light was cast on the mechanics of the general circulation by a series of laboratory experiments involving the rotation of shallow pans filled with water. The motion of the water was intended to model the motion of the atmosphere on a rotating earth, and to aid analysis a metallic powder was added as a tracer. To simulate the power source for the atmospheric circulation, the rim of the dishpan (that is, the 'equator') was heated, while the centre of the pan (the 'pole') was refrigerated. An 'equator' to 'pole' temperature gradient was thus created.

When the pan was rotated very slowly, a simple convective cell was observed to form, redistributing the heat from the rim towards the centre. This 'Hadley' cell extended close to the centre of the dishpan in a manner not unlike that postulated in Figure 3.14.

As the speed of rotation of the dishpan was increased to velocities comparable with the earth's rotation, this simple circulation was observed to break down. In its place a wavy pattern developed with large eddies embedded in great meandering streams, whose amplitude extended on occasion a substantial fraction of the distance from 'pole' to 'equator'. Furthermore, the increased deflective force arising from a faster rotation was seen to result in a predominantly west-to-east flow, girdling the 'earth' and particularly prominent in the 'middle latitudes' (Figure 3.16).

Figure 3.16   Dishpan experiments – patterns of motion on a hemispherical surface

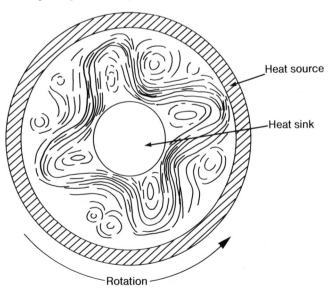

63

Figure 3.17 Balloon experiments – patterns of motion over southern temperate and polar latitudes. (Source: Gaskell and Morris, 1979)

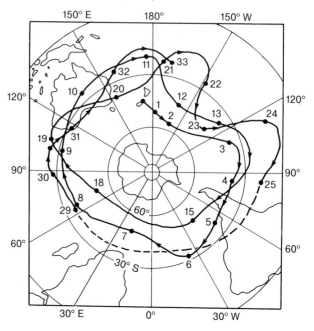

A striking comparison exists between the patterns displayed in the dishpan simulation model and actual atmospheric experiments, the results of one of which are shown in Figure 3.17. In this figure, the trajectory followed by a weather balloon launched from New Zealand was plotted as it was carried along by winds at a height of 12 000 m. Over the course of 33 days three circuits around the pole were completed along a meandering track in a west-to-east direction. Clearly, the undulating motions revealed in the dishpan experiments are major features of the general circulation and require further investigation.

(b)   The upper westerlies

The existence and intensity of the upper westerly flow is determined by the equator-to-pole temperature gradient. Figure 3.18 shows that pressure decreases more slowly with height in a column of warm tropical air than in a column of cold polar air, producing a pressure gradient from equator to pole. This gradient increases with increasing height. Air movement along it is modified by the Coriolis force to produce a westerly motion aloft, which characteristically exhibits a wavy pattern. Between two and five long waves may most commonly be observed in this flow. Normally they travel from west to east rather slowly, certainly much slower than the air blowing through them. Occasionally the waves may remain stationary for a time, or even retreat from east to west. Considerable variation is also evident in their north–south extent. At some times, the waves may be poorly developed (termed a *high zonal index circulation*), and

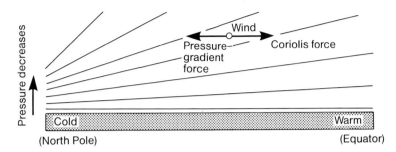

Figure 3.18   Cross-section showing the northward-directed pressure gradient responsible for generating the mid-latitude westerlies. (Source: Lutgens and Tarbuck, © 1979, p. 157. Adapted by permission of Prentice-Hall Inc., Englewood Cliffs, N.J.)

show hardly any north–south component. At other occasions, such extreme undulations are apparent that large cells of air may be detached from the mainstream of the circulation. Often these *low zonal index circulations* are associated with weather anomalies because of the manner in which they 'block' the normal west-to-east passage of weather systems at the surface. If the circulation becomes entrenched in this mode, the anomaly may prove quite persistent. The European droughts of 1975, 1976 and 1984, for example, were associated with such 'blocking'. Such evidently important links between the upper westerlies and surface weather features are considered further in Chapter 6.

The average height of the 500 mb surface for January and July is shown in Figure 3.19. At this altitude, wind flow is geostrophic and therefore parallel to the contours, with wind speed inversely proportional to the contour spacing. It is evident that, even averaged over a long time, a wave pattern exists. This is initially unexpected, since it would seem reasonable to expect that averaging out the waves would produce a circular contour pattern centred on the pole. The only explanation is that the troughs and ridges apparent on the averaged charts must represent preferred locations at which troughs and ridges are more frequently experienced than elsewhere. Two troughs are particularly prominent, one over north-eastern North America and the other over eastern Siberia. In both cases, an interaction between relief and the upper air circulation is primarily responsible. The disturbance of the westerly flow aloft by the Rocky Mountains and the Tibetan Plateau, both of which extend over 3 km into the troposphere over extensive areas, generates troughs in the flow downstream of these obstacles. These troughs are more marked on the winter chart (Figure 3.19 (*a*)), where they are further amplified by the coldness of these continental interiors. By contrast, weak ridges are apparent over the eastern Atlantic and Pacific Oceans, a consequence of their relative warmth during this season. The preferred positions of the upper westerly waves thus reflect on the thermal, as well as the mechanical (or dynamical), influences of the surface. Anchoring of moving waves, and the development of blocking situations, is more likely when waves are in these locations. The location and movement of these upper air features is important also because of their role in influencing the

Figure 3.19 Average height (in
metres) of the 500 mb surface
(*a*) in January; (*b*) in July.
(Source: Battan, 1979)

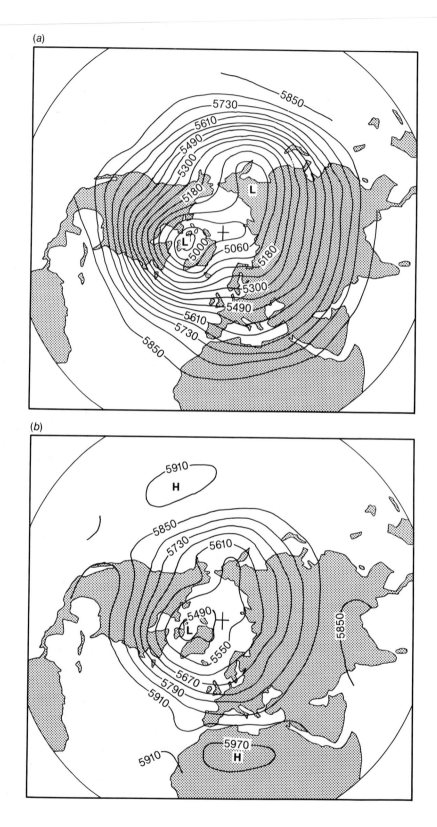

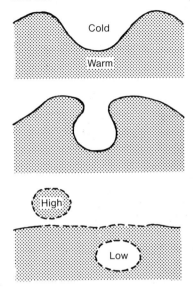

Figure 3.20 Eddy transfer of warm air polewards and cold air equatorwards

formation and direction of movement of surface weather systems. This is of vital importance for the regional climates of the middle latitudes, and will be considered further in Chapter 6.

The upper westerly waves perform one further fundamental function. These great horizontal waves, sometimes called *Rossby waves*, provide a mechanism for the transport of heat towards the poles, and thus for the balancing of the earth's energy budget. Figure 3.20 illustrates how these eddies can shift warm air bodies polewards and cold air equatorwards, by means of horizontal mixing. It was this possibility that the earlier generations of circulation models, with their emphasis on convective overturning, or vertical mixing, overlooked. Eddy motions, including travelling depressions and anticyclones, are now known to account for the bulk of heat transfer occurring outside the tropics. The current view of the general circulation therefore stresses the importance of heat redistribution predominantly via vertical mixing in low latitudes and via horizontal mixing as a result of eddy motions occurring in the upper westerlies over middle and high latitudes.

(c) Jet streams

During the Second World War bomber planes capable of flying at over 400 km/h found themselves making little headway on some missions involving east-to-west flight directions. By contrast, on their return flight, the time taken often indicated speeds over the ground of up to twice this velocity. The explanation lay in the existence of bands of strong westerly winds aloft, which have become known as the *jet streams*. These narrow cores of rapidly moving air commonly exhibit speeds of up to 250 km/h, compared with a typical westerly flow in which they are embedded of 50–100 km/h. Extending to widths of a few hundred kilometres, and with a vertical thickness of a few thousand metres, these fast-flowing rivers of air are usually to be found above heights of 10 000 m.

Like the upper westerlies, the jet streams are also a product of the all-important temperature gradient between the equator and the poles. The mechanism whereby this thermal contrast produces a pressure gradient and subsequently the geostrophic westerly winds aloft was illustrated in Figure 3.18. However, on closer analysis of actual temperature data, it is apparent that this thermal contrast is not a result of a regular temperature decrease towards higher latitudes. Rather, much of it is concentrated into a few narrow zones where cold and warm air masses come into contact. Such boundaries are known as *fronts* and often exhibit quite large horizontal differences in temperature over a relatively short distance on either side of the boundary. From Figure 3.21, it may be deduced that this intensification of the thermal gradient will produce an intensification of the induced pressure gradient aloft. Consequently, a marked strengthening of the upper westerlies may be expected to occur at such locations. These enhanced westerlies correspond to the jet streams. Like the westerlies, the jet stream will be best developed during winter, when the equator–pole temperature gradient is strongest.

Figure 3.21 Frontal zones and jet-stream formation. (Source: Trewartha and Horn, 1980)

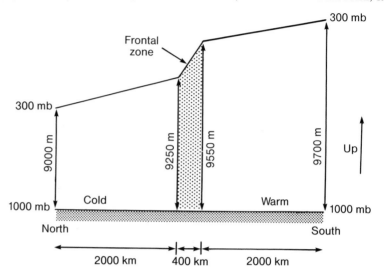

A particularly well-defined frontal zone exists at the poleward limit of the surface westerly circulation, where the tropical and polar air masses interact. The best-known of the jet streams, the *polar front jet*, is frequently identified above this frontal zone, meandering discontinuously from west to east. Because of the variability of its path, it does not stand out on charts of the mean circulation, although the close spacing of isolines on individual daily charts are tell-tale indicators of its location and intensity. Airline pilots plan their altitude and route on the basis of such knowledge, either to avoid or to ride these jet streams, depending on their flight direction.

Several other jet streams are known to exist, of which two are worth noting. The *subtropical jet* forms as a result of a frontal zone aloft at about 25° latitude, where the Hadley cell and mid-latitude circulations interact. This meanders much less than the polar jet and seldom exhibits the very high speeds of the latter. Secondly, when the zone of most intense heating (the intertropical convergence zone) has migrated north of the equator in the northern hemisphere summer, the upper air over West Africa and monsoon Asia is often warmer than that over the Gulf of Guinea and Indian Ocean respectively. In such circumstances, an *easterly jet* has been observed at heights of about 13 km, especially over southern India.

The westerly jets play an important role in governing short-term variations in the general circulation. First, they provide additional dynamism to the horizontal mixing role fulfilled by eddies in the upper westerlies. Since they occur at air mass boundaries, the meandering of the jets is an effective mixing mechanism stranding pools of warm or cold air in air bodies of considerably different temperature characteristics. Secondly, as will be shown later, the existence of these fast-flowing ribbons of air has significance for the formation and guiding of depressions and anticyclones at the surface. In stirring and mixing the air below by driving chains of swirling cyclonic storms north-eastwards, a further mechanism for balancing the earth's energy budget exists.

(d) Numerical models of the circulation

Circulation models, like all models, are imperfect simplifications of reality. As our knowledge of the functioning of the atmosphere has grown, the models invoked to aid our understanding have become progressively more elaborate. But, however complex they become, models inevitably stress certain relationships at the expense of others. Although this often results in a clearer definition of the most important components of the atmospheric circulation, some of the omissions may be serious. For example, the dishpan experiments made no mention of the terrestrial geography of the globe. The distribution of land and sea, the location of high mountain ranges, the effects of snow and ice at the surface are not considered because of the complexities that they introduce into the model. Such complexities, and many more, can now be incorporated into the latest generation of models, which make use of high-speed computers to consider the operation of many interacting variables in simulating the behaviour of the atmosphere.

The potential offered by this approach is very exciting. First, if the state of the atmosphere at a given time is known, the model can be run to provide a predictive simulation for a future time period. The availability of a greater network of world-wide data sources, particularly from satellites, brings this possibility closer. Second, the model can be run to incorporate slight internal changes. For example, what would happen if the level of carbon dioxide was doubled, or if the dust content of the atmosphere was increased, or if the tropical forests were denuded? Questions like these, of major importance for the future well-being of the global community, may be nearer resolution as a result of this modern approach to understanding and appreciating the workings of the atmospheric system.

ASSIGNMENTS
1. (a) *Explain what is meant by the terms 'Hadley cell' and 'indirect cell'.*
   (b) *Briefly summarise the limitations apparent in the unicell and tricellular circulation models.*
   (c) *What factors appear to be responsible for the general subsidence of air in the latitude zone 20°–35°?*
2. (a) *What are 'Rossby waves' and what is their significance to the general circulation?*
   (b) *What methods were employed in identifying their principal characteristics?*
   (c) *Suggest what a 500 mb average-height chart for January and July in the southern hemisphere would show in terms of preferred wave positions.*
   (d) *Explain how an easterly jet may develop at low latitudes during the hot season.*

3. *From your knowledge of the functioning of the general circulation, what changes would you envisage in the global circulation if any of the following occurred:*
   (a) *the Arctic Sea ice was deliberately melted to produce a warming up of the Arctic region by a few degrees centigrade;*
   (b) *the northern hemisphere was extensively glaciated;*
   (c) *a series of major volcanic eruptions occurred close to the equator?*

Key Ideas

A.   Importance of Air Movement

1. Horizontal air movement, or wind, is of major importance because of its ability to relocate heat and moisture at all areal scales. This process is known as advection.
2. The redistribution of heat energy from low-latitude areas of surplus to high-latitude areas of deficit is achieved mostly by wind action, enabling much more of the earth to be habitable than would otherwise be so.

B.   Forces that control Atmospheric Motion

1. Air motion is controlled by a mix of forces, including the pressure-gradient force, the force of gravity, the Coriolis force and frictional forces.
2. Air motion is initiated by a pressure gradient between places, with initial movement occurring from high to low-pressure locations.
3. A rapid reduction in air pressure with increasing altitude is apparent, from an average surface value of 1013.25 mb to about 10 mb at a height of 30 km.
4. Isobars are lines joining points that have identical air pressures, and may be used to identify pressure features and pressure gradients.
5. The height of a given isobaric surface may be mapped to enable pressure distributions at particular heights to be examined.
6. Because we view movement from a moving platform, an apparent deflection of moving objects due to the earth's rotation occurs. The Coriolis force causes moving objects to be deflected to the right of their line of motion in the northern hemisphere and to the left in the southern hemisphere.
7. The strength of the Coriolis force increases as the speed of the moving object increases, and it also increases with increasing latitude, being zero at the equator and a maximum at the poles.
8. Roughness of the earth's surface is responsible for exerting a frictional drag on wind layers close to the ground.

C. Scales of Atmospheric Motion

1. Air movement aloft is parallel to the isobars, reflecting a balance between the pressure gradient and Coriolis force. This is known as the geostrophic wind.
2. Frictional drag at lower elevations reduces the Coriolis force somewhat, enabling the pressure-gradient force to move air obliquely across the isobars.
3. Surface friction is associated with converging flows of air in low-pressure areas and diverging flows in highs.
4. Pressure differences leading to wind motions may be thermally induced in coastal regions.
5. Thermal contrasts in heating and cooling may on occasion produce upslope and downslope movement of air in areas of varied topography, and render some locations prone to a higher or lower incidence of frost and fog.
6. Pressure differences leading to air movement on a planetary scale may result from spatial differences in energy inputs. This may produce a latitudinal zonation of pressure and winds.
7. Wind-driven ocean currents act also as agents of redistribution for the earth's energy budget.

D. Modelling the General Circulation of the Atmosphere

1. Hadley's unicell model, although useful for low latitudes, involves an unbalanced surface easterly component.
2. Rossby's three-cell model, although adequate for surface winds, unsatisfactorily accounts for upper air movement in the middle latitudes.
3. Convective models alone are unable to explain the scale of heat transfer achieved between low and high latitudes.
4. Upper air movement is predominantly west to east in response to the equator-to-pole temperature gradient. These winds are referred to as the upper westerlies.
5. A major mechanism of heat transfer occurs in the upper westerlies by means of eddy motions driven by the Rossby waves.
6. The location, amplitude and speed of movement of waves in the upper westerlies strongly influence weather and climate at the surface in the middle latitudes.
7. Jet streams are a response to strong horizontal gradients of temperature and are particularly prominent over frontal zones. They usually occur above heights of 10 km.
8. Computer modelling of the general atmospheric circulation promises a quantitative predictive ability concerning the environmental impact of humans, both at the surface and on the higher levels of the atmosphere.

Additional Activities

1. Construct an isobaric chart, using the following observations of surface air pressure (mb), which were made simultaneously at a number of locations in Britain and Ireland.

Lerwick	1002.4	Stornoway	1002.0	Wick	1001.4
Benbecula	1002.2	Aberdeen	1000.0	Tiree	1001.3
Edinburgh	999.0	Malin Head	1001.0	Ayr	998.9
Newcastle	997.6	Belmullet	1002.3	Isle of Man	997.8
York	997.8	Dublin	999.6	Aberystwyth	1000.3
Chester	998.8	Lincoln	999.7	Luton	1000.7
Ipswich	1000.8	Valentia	1002.6	Exeter	1002.3
Poole	1002.2	Dover	1003.0	Penzance	1004.0

(a) Describe the pressure distribution shown in your chart.

(b) Suggest, giving reasons, where wind speeds are likely to be highest.

(c) Suggest, with reasons, the wind bearings which may be observed at the following places: Lerwick, Valentia, Ipswich, Isle of Man.

2. The distance between the isobaric surfaces at 1000 mb and 500 mb was measured for three columns of air one day in October over Britain. The values 5646 m, 5388 m and 5131 m were obtained. One of the air masses originated from the Azores subtropical anticyclone; one from the ocean around Iceland, which had blown directly towards Western Europe; and one from the same source, which had approached from the west after a long sea journey over the Atlantic. Identify, giving reasons, which one you think originated in each of these sources.

3. Table 3.3 shows a north-to-south section through central Europe from Narvik to Rome. The average height of the 500 mb surface for January and July is shown, and also a series of actual observations made on consecutive days in July 1981.

(a) Locate the places on a sketch map. Mark at each place the average heights of the 500 mb surface shown for January and July.

(b) Explain why the height of the 500 mb surface is lower in winter than in summer.

(c) Comment on the significance of the average gradient in this isobaric surface for the strength of the upper westerlies.

Table 3.3   Heights of the 500 mb surface in central Europe for a selected sequence of days in July 1981

Station	Mean height of 500 mb (m)		Height of 500 mb (m)						
	Jan	July	3/7	5/7	7/7	9/7	11/7	13/7	15/7
Narvik	5257	5578	5370	5480	5590	5640	5680	5600	5520
Oslo	5360	5639	5520	5560	5660	5800	5760	5580	5520
Copenhagen	5425	5672	5600	5640	5710	5810	5780	5600	5590
Frankfurt	5486	5745	5730	5740	5800	5820	5800	5750	5680
Munich	5505	5760	5770	5800	5800	5830	5790	5760	5760
Milan	5532	5821	5830	5820	5820	5830	5780	5800	5840
Rome	5562	5852	5880	5870	5820	5830	5800	5780	5830

(d) For the period 3–15 July, note on which days the observed height is greater than average for each place.

(e) In what respects do your results indicate (i) the poleward movement of a warm air mass; (ii) the possible passage of a wave in the upper westerlies?

(f) By identifying where the maximum gradient in the 500 mb surface occurs, suggest a probable location for the polar jet on 9 July.

4 Atmospheric Moisture

Moisture, after energy and motion, forms a third key element of the atmosphere. Atmospheric moisture is important in weather and climate for three reasons. First, there is a close interdependence between atmospheric energy, motion and moisture. As shown in this chapter, the capacity of the air for vertical movement is influenced not only by its temperature and density, but also by its moisture content. In addition, the earth's major wind systems transport moisture as well as heat across the globe. Second, moisture in the air in its liquid and solid states as fog and cloud (this chapter) and as dew, rain, snow and hail (Chapter 5) provides a visible physical expression of the workings of the atmosphere. Third, in common with energy, a source of moisture is a fundamental requirement of all living systems. The detailed character of the planet's ecosystems is strongly affected by the distribution of available moisture at the earth's surface. An excess or, more commonly, a deficiency of available surface water can have a devastating effect on plants, animals and humans, as the recent droughts in the Sahel of sub-Saharan Africa so vividly demonstrate (Chapter 6).

A. The Hydrological Cycle

A convenient starting point in any analysis of atmospheric moisture is the concept of the *hydrological cycle*. This refers to the global circulation of moisture (and heat) between the land and sea surface and the atmosphere. The hydrological cycle, as shown in Figure 4.1, is composed of a series of stores or compartments in which moisture is held in various forms and amounts, and a sequence of transfers and transformations of moisture between and within the different stores.

1. Stores within the cycle

The moisture-storage capacity of the various compartments of the cycle indicated in Figure 4.1 is very uneven. The oceans hold the vast bulk of all global moisture (97%). Of all *fresh* water, 75% is in the form of ice-sheets and glaciers, and almost all the remainder is ground water. It is remarkable that rivers and lakes hold only 0.33% of total fresh water. Although it only has about 0.06%, the soil holds, at any one time, about twice as much moisture as the atmosphere, which has a mere 0.035%.

Figure 4.1 The hydrological cycle showing moisture stores and linkages of moisture between stores. Note that the circulation of moisture from land and ocean to the atmosphere can be seen as separate components. Moisture may be cycled *within* the ocean *or* land part of the system or *between* land and ocean compartments.

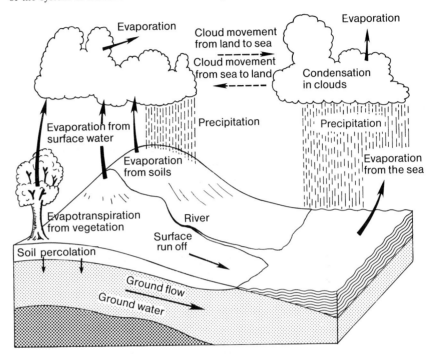

(a) Atmospheric store

Although the atmosphere represents a relatively small store, its moisture content is highly variable in space and time. It is also quite different from the other main stores, which tend to be dominated by liquid water and ice. Moisture exists in the air largely in gaseous form as water vapour. Only about 4% is in the form of liquid water droplets and ice crystals, found mostly in clouds.

The ability of the air to hold water vapour depends solely on temperature. A simple rule states that the warmer the air, the more moisture it can hold. When a mass of air is holding the maximum amount of water vapour possible at a given temperature (for example, the dew-point temperature), the air is said to be *saturated*. When it is retaining less than the saturated amount, the air is referred to as *unsaturated* and when holding more (under special circumstances), it is known as *supersaturated*. The moisture content of an air parcel, whether at saturation or not, can be described in terms of its specific and relative humidity.

Specific humidity is the ratio of the weight of water vapour held to the total weight of moist air. It is often expressed as the number of grams of water vapour per kilogram of moist air. As previously suggested, and as shown in Figure 4.2, the specific humidity of air at saturation (the *saturation specific humidity*) is greater in warm than in cold air. Thus at a temperature of 20°C a saturated air mass (at A_1) can hold about 15 grams

Figure 4.2 The mass weight of water vapour held by saturated air at different temperatures

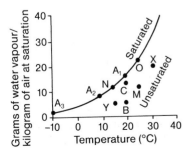

75

of water vapour per kilogram of air. At 10°C, however, saturation occurs with only half that amount (at A_2), and at –10°C a mere 1.5 grams of water vapour is sufficient to produce saturation. In addition, the specific humidity of an unsaturated sample of air at B is only 5 grams, but rises to 15 grams in the saturated sample at A_1. It is worth noting that the only way that the specific humidity of a parcel of air can be increased or decreased is to add or subtract water vapour. This is not the case when air humidity is measured in relative terms.

Relative humidity is the percentage ratio between the actual specific humidity and the saturated specific humidity, that is:

$$\text{Relative humidity} = \frac{\text{actual specific humidity}}{\text{saturated specific humidity}} \times 100$$

The relative humidity of the air sample at B is only 33%, since it is retaining 5 grams out of a possible 15 grams (at saturation) of water vapour per kilogram of moist air. This is in contrast to the air sample at C which, since it holds 12 grams of water vapour out of a possible 15 grams, has a relative humidity of 80%.

The relative humidity is thus a reflection of how far the air is to saturation. Low relative humidity values of 20–30% mean that the air is 'dry' and a long way from saturation. Higher percentage figures, for example 80–90%, signify that the air is 'moist', and when full saturation is reached the relative humidity is 100%. Because the relative humidity depends not only on the specific humidity, but also on actual air temperature (which controls the saturated specific humidity), relative humidity is a highly variable measure.

2. Linkages within the cycle

Figure 4.1 illustrates the main linkages or exchanges of moisture between the different stores. Since the average moisture content of the main compartment is neither increasing nor decreasing over time, there is clearly an overall balance between the exchanges. The various moisture linkages are composed, first, of a number of phase changes of moisture including that from a liquid to a gaseous state (evaporation), and from a gaseous to a liquid state (condensation). Second, there are vertical transfers of moisture between stores, including downward precipitation and the upward transfer of evaporated water. Finally, there are lateral transports of moisture between and within stores by wind movement and surface runoff.

(a) *Phase changes of moisture*

During the process of *evaporation*, water, a liquid with closely packed molecules, is transformed into water vapour, a gas with widely separated unconstrained molecules. As indicated in Figure 4.3, a large amount of energy (about 600 calories) is required for every gram of water so evaporated. This energy is generally provided by the removal of heat from the immediate environment in which evaporation is taking place, with a con-

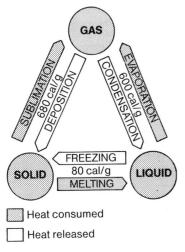

Figure 4.3 Phase changes of moisture. (Source: Trewartha and Horn, 1980, p. 43)

▨ Heat consumed

▢ Heat released

sequent reduction in local temperatures. Because the energy involved in evaporation changes the state of moisture from a liquid to a gas *at the same temperature*, such energy is called hidden or *latent heat*.

Condensation is the phase change of moisture from a gaseous to a liquid state, and latent heat (600 calories per gram) is normally released into an environment where condensation is occurring. Nevertheless, as shown in section B below, condensation usually takes place when air is chilled. If an air parcel is cooled to saturation (that is, to its dew-point temperature), gaseous moisture (water vapour) within the air begins to condense out into tiny water droplets. In Figure 4.1, warm moist air charged with water vapour rises and cools at altitude. When the dew-point temperature is reached, condensation occurs and clouds form. These clouds of course may evaporate again if temperatures rise, but they may ultimately lead to precipitation.

It may be noted that there are a number of other phase changes of moisture which take place within the hydrological cycle (Figure 4.3). *Melting*, the change from a solid to a liquid form, consumes about 80 calories per gram and produces a local cooling response, whereas *freezing*, the change from a liquid to a solid state, releases an equivalent amount of heat into the environment.

(b) Transfers: vertical and lateral movements

As shown in Figure 4.1, precipitation (rain, snow, hail) is the main method by which moisture is removed downwards from the atmosphere to enter the land/ocean part of the cycle. Conversely, the upward transfer of evaporated moisture is the principal way moisture enters the atmosphere from the land and sea surface. About four-fifths of the moisture input to the atmosphere is from evaporation over the oceans: the remaining one-fifth originates from evaporation over the land surface. The latter includes evaporation losses from inland bodies of water, moist soil and from water intercepted by vegetation. Moisture removal from the land surface also includes that lost from the soil through the stems and leaves of plants to the air in the process of *transpiration*. The term *evapotranspiration* is often employed to denote the sum total of moisture removed by evaporation and transpiration from a vegetated land surface.

Transfer linkages within the hydrological cycle are completed by a series of horizontal shifts of moisture. Although there are significant lateral transfers of moisture within the main stores, there is a net transport of atmospheric moisture by wind movement from the oceans to the continents. To counter this effect, there is an equivalent return flow of water from the land to the sea by *surface runoff*.

The total moisture exchange between the stores of the hydrological cycle is considerable. The air, for instance, contains, without recycling, sufficient moisture (2.5 cm of water-equivalent depth) for only about 10 days' supply of precipitation for the earth as a whole. The average global precipitation rate is 85.7 cm per year, however, and this implies that there is quite substantive recycling between land, ocean and atmosphere.

ASSIGNMENTS

1. (a) Explain what is meant by the concept of the hydrological cycle.
   (b) Describe the main stores and linkages of moisture within the cycle.
   (c) With reference to Figure 4.1, complete the model of the hydrological cycle shown in Figure 4.4, using the following key words: ground water flow, evapotranspiration, infiltration, vegetation, evaporation, precipitation, surface runoff, evaporation from soil.
   (d) Which of the two models shown in Figure 4.1 and 4.4 do you prefer and why?

Figure 4.4   Flow diagram of the hydrological cycle (see assignment A1)

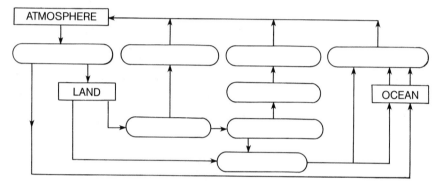

2. Refer to Figures 4.2 and 4.3.
   (a) Describe the various states of moisture in the atmosphere and the processes of phase change.
   (b) Explain what is meant by saturated and unsaturated air.
   (c) Define the terms specific humidity and relative humidity.
   (d) Using Figure 4.2, describe the relationship between temperature and the specific humidity of air at saturation.
   (e) Calculate the specific and relative humidities of the air samples at X and Y.
   (f) Using your results, explain why relative humidity is a much more variable element than specific humidity.
   (g) Name two ways in which the unsaturated air sample at M can be increased to saturation (that is, at N and O).

B. Condensation: Basic Mechanisms

Condensation is an important atmospheric process, since it is responsible for cloud and fog and performs a vital role in the formation of precipitation. Condensation takes place when air is brought to saturation. The condensation process is supported in the atmosphere by the presence of atmospheric nuclei.

1. Air cooling

The most common way that air is brought to saturation and condensation is induced is by air cooling. As indicated in Figure 4.2, a sample of air at M with a relative humidity of 50%, a specific humidity of about 10 grams of water vapour per kilogram of air and a temperature of 26 °C may be induced towards saturation by cooling along the line MN. The cooled air sample at N has a relative humidity of 100%, a specific humidity of 10 grams of water vapour per kilogram of air and a temperature of 14.5 °C.

(a) The causes of cooling

Apart from *air mixing*, when a warm mass of air can be cooled when intermixed with a cold body of air, air cooling can take place in the atmosphere in three main ways. First, it can be cooled by *advection* (or horizontal movement), when warm, moist air is cooled from below as it passes over a cold land or sea surface. The second method of cooling air is by strong *radiation*, when a land surface chilled by outgoing radiation under clear skies in turn cools the air above. As these two processes and the forms of condensation associated with them commonly occur in very stable air, they will be examined in section D below. The most effective way air can undergo a reduction of temperature, however, to produce saturation and condensation is by *vertical ascent*.

Figure 4.5 The effect of orographic and frontal barriers, convergence and convection on air uplift

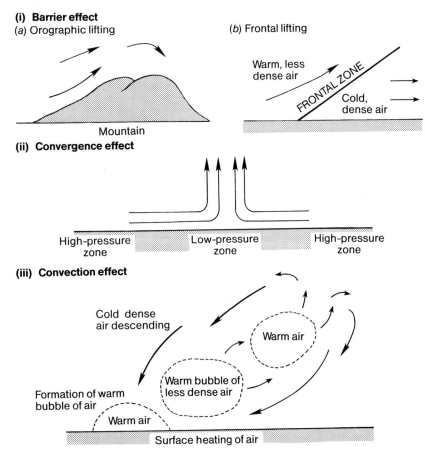

(i) Barrier effect
(a) Orographic lifting

(b) Frontal lifting

Warm, less dense air

FRONTAL ZONE

Cold, dense air

Mountain

(ii) Convergence effect

High-pressure zone Low-pressure zone High-pressure zone

(iii) Convection effect

Cold dense air descending

Warm air

Formation of warm bubble of air

Warm bubble of less dense air

Warm air

Surface heating of air

(b) Reasons for air uplift

Air may be forced to rise in response to three basic controls. First, horizontally moving air (that is, wind) may be forced to rise, if it encounters an obstacle in its path. *Orographic lifting* (see Figure 4.5) results when wind encounters hills and mountains, and perhaps even some cities and large vegetation formations. *Frontal uplift*, on the other hand, takes place when warm air is forced to glide over colder air in frontal zones. Second, large-scale uplift of the atmosphere also takes place when air *converges* or moves into low-pressure cells (see Figure 7.6). Third, small-scale *convective* currents cause vertical motion when the air is heated from below. As it is heated, warm air near the surface expands and becomes less dense. This warmer, less dense air rises through the atmosphere to be replaced by colder, denser air from above.

2. Atmospheric nuclei

Laboratory tests have shown that condensation can only take place in the atmosphere, if there are particles or impurities around which liquid water can form. The atmosphere contains such *condensation nuclei* in the form of microscopic particles of salt, dust and smoke.

(a) Condensation and hygroscopic nuclei

Although condensation nuclei are generally abundant in the atmosphere, they vary greatly in size, concentration and chemical composition from one area to another. Very high concentrations of sulphur dioxide and smoke may be found in the atmosphere above cities and industrial regions from fossil fuel-burning, and over rural areas the concentration of fine dust particles from wind-blown soil may also be quite high. Some substances, such as sulphur dioxide and salt, have a very strong affinity for water; these nuclei are very powerful and permit condensation to occur below 100% relative humidity, even as low as 75%. They are referred to as *hygroscopic nuclei*.

Once condensation has begun on a nucleus, the droplets grow rapidly, attaining sizes of up to 0.10 mm (100 μm) diameter, which is the average size of most cloud droplets. Such droplets are little affected by the earth's gravity and remain suspended in the atmosphere, collectively making up clouds and fog. The crucial formation of the larger water drops necessary for precipitation requires special processes in addition to condensation; these are dealt with in Chapter 5.

(b) Freezing nuclei and supercooled water

The water droplets of clouds and fog are very difficult to freeze, even at low temperatures below 0°C. Indeed liquid water droplets may exist in clouds as supercooled droplets in temperatures as low as −40°C. This is because the formation of ice crystals, an important constituent of many clouds, needs particular nuclei called *freezing nuclei*, in addition to very low temperatures. These nuclei, which are most active or effective between −20°C and −25°C, are much less abundant than condensation nuclei. Certain types of soil particles (clay), dust and even meteoric material may act as nuclei for the freezing mechanism. We are thus led to conclude that clouds and fogs contain any or all of the following: water droplets with temperatures greater than 0°C, supercooled water droplets with temperatures between −40°C and 0°C, and ice crystals, especially when temperatures are lower than about −20°C.

ASSIGNMENT
1. (a) *What is the main method by which air can be brought to saturation and condensation induced?*
   (b) *Give four different ways in which air can be reduced in temperature.*
   (c) *List three causes of vertical air ascent.*
   (d) *Examine the role of atmospheric nuclei in cloud and fog formation.*

C. Condensation: Subsequent Development

The subsequent flight path of a parcel of air forced to ascend through the general atmosphere by orographic, frontal, convergent or convectional effects depends on the temperature, and thus the density of that parcel in relation to its surroundings. If an air parcel finds itself warmer and thus less dense than its environment, it will be encouraged to rise still further. If it is cooler and therefore more dense, it will be forced to continue descent. In order to comprehend the forces which determine the buoyancy of an air parcel in the atmosphere, and thus the likelihood of condensation and cloud formation occurring from it, it is necessary to examine three ideas. First is the concept of temperature lapse rate; second the process of adiabatic temperature change; and third the subject of air stability.

1. Lapse rates and adiabatic processes

(a) *Environmental lapse rate*

The actual change of temperature with height in the atmosphere, such as that experienced by a thermometer moving vertically through the atmosphere, is referred to as the *environmental lapse rate*. Figure 4.6 indicates that the environmental lapse rate (ELR) can vary considerably in time and space, depending on local air temperature conditions, but the average ELR shows a decrease of temperature of about 6.5°C per kilometre.

(b) *Adiabatic processes and lapse rates*

If a parcel of air is forced to move vertically through the atmosphere (by one or a combination of the mechanisms of uplift indicated in section B above), it will experience a number of changes. As pressure decreases

Figure 4.6 Lapse rates of temperature with height: ELR is the environmental lapse rate and can be (i) steep, (ii) gentle, or (iii) show an increase of temperature with altitude; SALR is the saturated adiabatic lapse rate; and DALR is the unsaturated or dry adiabatic lapse rate

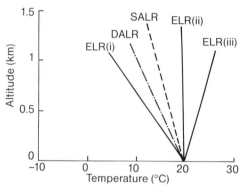

with altitude, a parcel of rising air will expand and consequently cool. A parcel of sinking air will be compressed and warmed. Now, if the moving parcel of air ascends (or descends) fast enough, very little heat will be exchanged between the parcel and the surrounding air. If heat is neither added to, nor removed from a moving air parcel by the external environment, the process is called *adiabatic*. Under adiabatic control, all temperature changes within the air parcel are due to expansion or contraction. Under such circumstances, a rising air parcel is subjected to adiabatic cooling and a descending parcel to adiabatic warming. The rate at which temperature changes in a rising, expanding (or a sinking, compressed) parcel of air is called the *adiabatic lapse rate*. There are two different types of adiabatic lapse rates, depending on the moisture status of the air parcel.

(i) *Dry adiabatic lapse rate (DALR)*. If a rising (or falling) parcel of air remains dry – that is, unsaturated – with a relative humidity below 100%, it will decrease (or increase) in temperature at a rate of about 10°C per kilometre. This is known as the *dry adiabatic lapse rate* (DALR) (see Figure 4.6).

(ii) *Saturated adiabatic lapse rate (SALR)*. If a rising parcel of dry air continues to cool and reaches saturation, latent heat is liberated as shown in Figures 4.6 and 4.7, which counteracts somewhat the dry adiabatic temperature decrease. Similarly, a descending mass of saturated air will warm less quickly than dry air, because latent heat is absorbed to evaporate moisture within the air parcel. The lower rate of temperature change in a moving saturated parcel of air is called the wet or *saturated adiabatic lapse rate* (SALR). The SALR is normally about 5–6°C per kilometre in the warmer, moister air of the lower atmosphere, but may increase to near the dry adiabatic rate in the cold, dry air of the upper troposphere (see Figure 4.8).

(iii) *Condensation level*. A rising, expanding parcel of dry air will continue to cool at the DALR, until it reaches its dew-point temperature and saturation occurs. Thereafter the saturated air mass will cool at the

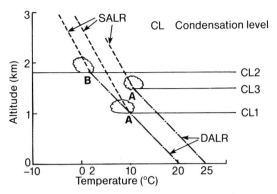

Figure 4.7 Influence of initial air temperature and the dew-point temperature on the condensation level, that is, the height at which clouds form. Dew point temperature of air parcel at A is 10°C and at B 2°C

SALR. The height at which condensation takes place is referred to as the *condensation level* and also, since cloud begins to form at this point, as the *cloud base*. As shown in Figure 4.7, the height of the cloud base is a function of both the dew-point temperature of the rising air mass, and the temperature at which the air mass begins its first ascent. An air parcel with a temperature of 20 °C and a dew-point temperature of 20 °C will condense at a much higher level (CL2 at B) than one with a dew point of 10 °C (CL1 at A). Similarly, the height at which clouds form will fall from level 3 to level 1 if the initial temperature of the rising air parcel is reduced from 25 °C to 20 °C.

2. Stability

The relationship between the environmental lapse rate (ELR) and the dry and saturated lapse rates (DALR, SALR) determines the stability of the atmosphere at any particular place. Air stability is a very important meteorological phenomenon, because it influences the amount and the type of condensation (clouds, fog) which take place, together with other related weather phenomena, such as rain and hail.

(a) Unstable air

A condition of *instability* exists when uplifted air is encouraged to rise still further and descending air to sink. As shown in Figure 4.8, instability occurs when the environmental lapse rate is greater than that of either the wet (BC) or dry (AB) adiabatic lapse rates. Under such circumstances, a rising air parcel following path curve ABC will become progressively warmer and a sinking parcel (CBA) progressively cooler compared with the surrounding air. Both situations will promote further displacement of air, upwards and downwards respectively, from its original position. Unstable air conditions, especially if they prevail for several kilometres through the atmosphere, are closely associated with the towering cumulus types of cloud (for example, cumulonimbus) which may give rise to heavy thunderstorms (see Chapter 5, section C).

Figure 4.8 Unstable air conditions. Here the environmental lapse rate (ELR 1 and lower part of ELR 2) is greater than both the dry and saturated adiabatic lapse rates. Associated weather conditions are also shown.

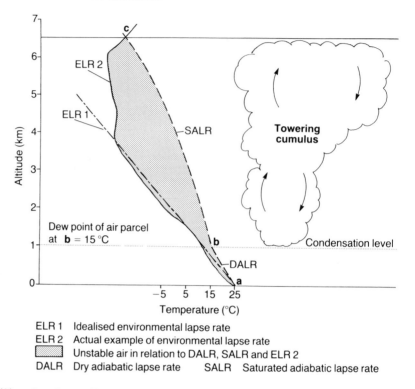

ELR 1   Idealised environmental lapse rate
ELR 2   Actual example of environmental lapse rate
    Unstable air in relation to DALR, SALR and ELR 2
DALR   Dry adiabatic lapse rate   SALR   Saturated adiabatic lapse rate

(b) Conditionally stable air

Conditional stability occurs when the environmental lapse rate is less than the dry, but greater than the saturated adiabatic lapse rate. Figure 4.9 shows that this type of stability occurs when moist air is forced upwards

Figure 4.9 Conditional stability. This occurs when the environmental lapse rate (ELR 1 and lower part of ELR 2) is less than the dry adiabatic lapse rate but greater than the saturated adiabatic lapse rate.

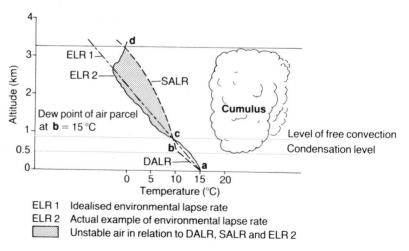

ELR 1   Idealised environmental lapse rate
ELR 2   Actual example of environmental lapse rate
    Unstable air in relation to DALR, SALR and ELR 2

(ABCD) and is at first cooler than its surroundings. At some point during the ascent (at B), condensation will take place and latent heat will be released into the rising air parcel. Cooling less rapidly at the SALR, it will ultimately become warmer than the surrounding air (at C). At this stage, it will become unstable and will continue to rise under its own buoyancy, until once again it reaches the temperature of its surrounding environment.

(*c*) *Stable air*

A state of *stability* is said to exist in the atmosphere when a vertically displaced parcel of air tends to return to its original position. This condition is shown in Figure 4.6 and occurs when the environmental lapse rate (ELR(i) and ELR(ii)) is less than the dry and saturated adiabatic lapse rates. In this case, the rising, adiabatically cooling parcel of air will be at a lower temperature than its surroundings, whereas a descending, adiabatically warmed parcel will be warmer. In each case, the parcel of air will tend to return to its original position by moving downwards in the first case and upwards in the second.

Under stable air conditions, a parcel of air forced to rise is unlikely to go on rising. It may, however, cool to dew point and, before descending, give rise to thin stratus (horizontal) or fair-weather cumulus clouds.

Figure 4.10 Stable air conditions and associated cloud development

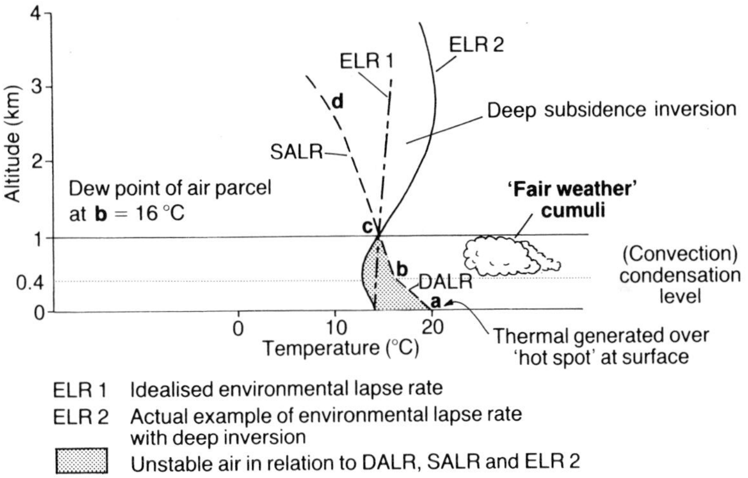
Altitude (km)
4
3
2
1
0.4
0
ELR 1
ELR 2
d
SALR
Deep subsidence inversion
Dew point of air parcel
at **b** = 16 °C
'Fair weather' cumuli
c
b
DALR
a
(Convection) condensation level
0
10
20
Temperature (°C)
Thermal generated over 'hot spot' at surface
ELR 1 Idealised environmental lapse rate
ELR 2 Actual example of environmental lapse rate with deep inversion
Unstable air in relation to DALR, SALR and ELR 2

Figure 4.10 shows how 'fair weather' cumuli clouds also form when there is instability in the air close to the ground surface but air stability above. In this example, summer day time heating produces local cells of warmed air (called thermals) at the surface. These hot air cells are unstable following the path curve ABC with condensation taking place at B. The 'fair weather' cumuli which form are capped at C because of a deep subsidence inversion with warmer air temperature above.

(*d*) *Highly stable air*

When, as indicated by line ELR (iii) in Figure 4.6, air temperature increases with altitude, a condition known as a *temperature inversion* exists. Temperature inversions are shown at different levels in Figure 4.9 (about 2.5–3.5 km) and in Figure 4.10 (about 0.5–2.5 km). Temperature inversions produce highly stable atmospheric conditions, and greatly curtail the vertical ascent of air below them. They effectively put a cap on the atmosphere. For instance, if an air parcel at sea level is forced to rise either at the DALR or the SALR, it will always be colder than the surrounding air which is increasing in temperature with height. Being colder than its environment, the air parcel will sink back to ground level and little cloud development will take place.

ASSIGNMENTS
1. *Refer to Figures 4.6 and 4.7.*
    (a) *Describe the nature of the environmental lapse rate (ELR).*
    (b) *Explain its formation and suggest why it varies in time and space.*
    (c) *What is meant by adiabatic temperature change?*
    (d) *Examine the dry adiabatic lapse rate (DALR) and the saturated adiabatic lapse rate (SALR).*
    (e) *Why do clouds begin to form (that is, at the cloud base) at different levels in the atmosphere?*
2. *Refer to Figures 4.8–4.10 and the text.*
    (a) *Describe the four states of air stability.*
    (b) *Explain the development of the different types of air stability.*
    (c) *Examine the links between air-stability conditions and cloud formation.*

D. Forms of Condensation

There are two main forms of condensation, based on their origin, extent and position in the atmosphere. Minor forms include fogs, mostly at low levels at or near the earth's surface, and are associated with very stable air conditions. Clouds represent the principal and most extensive form of condensation. The majority of clouds form at a distance from the ground surface and are related to rising currents of air, especially in more unstable atmospheric conditions.

1. Fog

(a) Advection fog

This takes place when air in contact with the earth's surface flows from a warmer to a colder area. If the air in contact with the cold surface cools to dew point, advection fog or low *stratus* cloud results (see Plate 4.1). This happens, for instance, on the western coastal margins of Europe in winter, when a mild, moist south-westerly flow of air from the Atlantic passes over a colder land surface. Fog and low stratus cloud also form in spring and summer over a cool North Sea, when there is an easterly or south-easterly flow of warm, relatively dry air from the continent of Europe. After initially picking up moisture (by evaporation) in its passage

Plate 4.1 NOAA–7 visible light image, 1414 GMT, 9 July 1983, showing classic advection fog conditions in the North Sea and adjacent land areas. An anticyclone lying to the north of the region set up a weak, but significant easterly flow of air on 9 July. This circulation, enhanced by local sea breezes along the east coast of Great Britain, allowed warm air from the continent (average daytime land temperatures reached 28°C over Holland) to be chilled to dew point as it passed across a cool North Sea (average July temperature equalled 14.5°C). The high ground over the North Yorkshire Moors stands above the fog (haar) and suggests a depth of around 300 metres or less. (Photograph: University of Dundee)

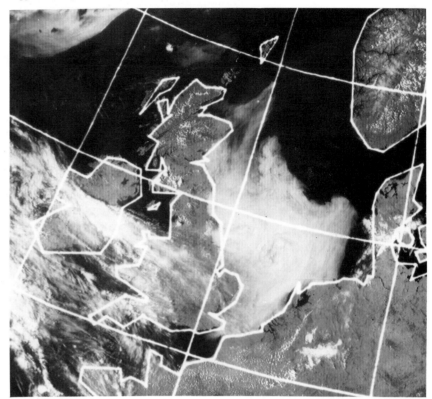

over the North Sea, this air stream subsequently cools and condensation occurs. Such processes give rise to the North Sea haars along the eastern coast of Britain. These haars, however, quickly dissipate a few miles inland over the much warmer land surface.

(b) Radiation fog

Radiation fog results when a mass of moist air, in contact with a cooling land surface, is chilled to dew point. Such fog is particularly related to the development of nocturnal ground-based temperature inversions (see page 28). Although highly stable air conditions are necessary for the formation of temperature inversions at ground level, and for the creation of radiation fog, a light wind of 3–10 km/h is required to produce fog. This is because a relatively large amount of moisture is needed to come into contact with a cooling land surface, a circumstance which cannot be achieved if the air is stagnant. With a gentle wind, sufficient turbulence is produced to mix the air chilled by ground contact with some of the warmer, moist air aloft. Higher wind speeds, on the other hand, tend to prevent cooling to saturation and the formation of fog, since too much warm air from above is brought down to the surface, raising overall temperatures above the dew point.

Once a layer of fog has formed with a depth of between 3 and 350 m, continued radiation from the top of the fog itself (and not from the ground) gives rise to the development of a temperature inversion immediately above the fog layer. This stabilises the fog and can help to maintain it for some time.

(i) *Valley fog.* Radiation fog is amplified in valley bottoms. Air chilled by rapid outgoing radiation from over surrounding hill land increases in density and sinks down adjacent valley slopes. Such cold air drainage quickly leads to a temperature inversion, where very cold air in the valley bottom is overcapped by warmer air above. The degree and speed with which air is cooled within the valley and the stability of the air under the temperature inversion (reinforced by the confines of the slopes) encourages the rapid cooling of air to dew point and the formation of valley fog.

(ii) *Industrial fog (smog).* Radiation fogs, whether developing in valleys or on flat terrain, tend to be intensified in the vicinity of cities and industrial areas. In these areas, air pollution provides additional condensation nuclei such as smoke and sulphur dioxide for fog formation. When pollution (smoke) is mixed at high density with fog, thick smog (smoke + fog) develops, and may prevent a substantial amount of radiation from reaching and warming the ground, a situation which helps to perpetuate the fog event.

On the whole, unless they are very thick, fogs do not persist for more than a few days. Fog dispersal results either from a sufficient intensity of daytime insolation to cause evaporation, or an increase of wind speed to mix the fog layer with the drier and usually warmer air above.

2. Cloud formation and classification

Clouds have been referred to as the 'writings of the atmosphere', for they indicate subtle processes in the vertical exchange of heat and moisture between the earth and the atmosphere. Accordingly, a grouping of clouds, based on their overall form, composition and height in the atmosphere goes beyond mere description. Such a classification gives clues to the dynamics of the weather.

(a) Stratiform clouds

Stratiform clouds, as the name implies, exhibit stratification. They have an extensive and usually a flat or sheet-like appearance. Stratiform clouds often develop due to the forced ascent of air under 'stable atmospheric conditions' (see Figure 4.10). The agency for ascent is often large-scale air flow towards the centre of a low-pressure system. Extensive forced ascent of the atmosphere leads to the formation of cloud sheets spread

Plate 4.2 Fair-weather cumulus over Fareham, Hants. Note the clear-cut horizontal bases of these clouds (at about 1000 m) and the flattened or slightly rounded tops. (Photograph: C. S. Broomfield)

over an area that is large horizontally, compared to its vertical extent (between 100 : 1 and 1000 : 1).

When such clouds appear as thin stratified layers below 1 or 2 km, they are called *stratus*. Sheet-like clouds at altitudes of between 3 and 4 km are called *altostratus*. A thick, dark stratified cloud yielding rain or snow is called *nimbostratus*.

Plate 4.3 A cumulonimbus of large vertical extent (A), developing from cumulus congestus (B). The tops of these massive clouds show a structure that often resembles an anvil plume (C). Squalls, hail and/or thunder often accompany the cumulonimbus cloud type. Individual cumulus clouds can be seen at D. Photograph taken at Beira, Mozambique, 2 April 1966 (Photograph: Crown Copyright. Reproduced with the permission of the Controller of Her Majesty's Stationery Office)

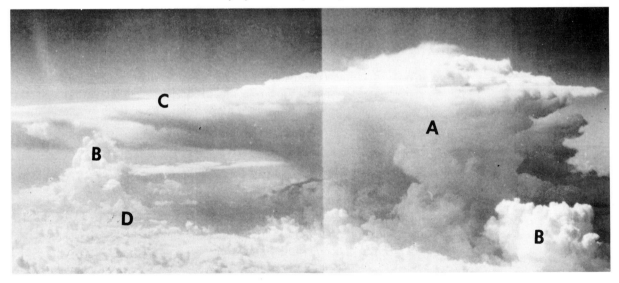

(b) Cumuliform clouds

An entirely different type of cloud, the cumuliform variety, forms mainly during buoyant air ascent in 'unstable air conditions' (see Figure 4.8). *Cumulus* are generally individual clouds, and have the appearance of rising mounds, domes and towers. The white cauliflower nature of the upper parts of these clouds signifies the buoyancy of cooling, condensing parcels of air.

Most cumulus cloud is of the fair-weather variety (Plate 4.2). Such forms may develop due to forced ascent in stable air (Figure 4.10). They may also occur, however, like other and larger cumulus forms, during the creation of localised instability during the day from local heating of the ground surface (Figure 4.8). Fair-weather cumulus are usually forced to descend and to vaporise, near sunset, because of increased stability brought about by evening cooling.

If atmospheric conditions are particularly unstable, cumulus may enlarge, first, into *cumulus congestus* and, finally, into *cumulonimbus*. These latter clouds are heavy masses of cloud with great vertical extent. Their mountainous summits may reach 10–15 km. An anvil-shaped top is characteristic, as the upper layers of the cloud tend to sprawl out horizontally below the high inversion at the top of the troposphere (Plate 4.3). They are often accompanied by sharp showers, squalls, thunderstorms and sometimes hail. Cumulonimbus are usually restricted to areas where there is both large-scale convergence of moist, warm air at low elevations and air instability, introduced by daily solar heating of the land surface.

When many cumuliform cloud elements are arranged in a layer, and the cloud is below about 3 km, with the elements appearing fairly large, the composite is called a *stratocumulus*. If the cloud layer is between about 3 and 6 km, with the individual elements appearing relatively small, the cloud is called *altocumulus*.

(c) Cirrus clouds

Cirrus types of cloud are composed mostly of ice crystals and occur at high elevations. These clouds are often formed of white, feather-like elements. When a cirriform cloud appears as a whitish veil which is fibrous or smooth and covers much or all of the sky, it is called *cirrostratus*. Higher layers of cloud composed of many small cloud elements arranged in groups, lines or ripples are referred to as *cirrocumulus*. As these clouds sometimes resemble the scales of a mackerel, the phrase 'mackerel sky' is used to describe them.

ASSIGNMENTS
1. (a) *Examine the atmospheric conditions which favour the development of fog.*
    (b) *Why are radiation fogs particularly associated with valleys and urban/industrial areas?*
2. (a) *Suggest a simple classification of cloud types.*
    (b) *Explain the development of cumulonimbus.*
    (c) *Describe and account for the weather associated with it.*

E. Orographic Effects

It is possible to summarise some of the heat, motion and moisture relationships outlined in this chapter. This can be pursued by examining a range of effects that orographic barriers have on the flow of moist air streams.

1. Lee waves and lenticular clouds

When an air stream is forced to rise, due to an obstacle such as a mountain range, the relief feature may throw the air stream into a series of waves above and downstream of the barrier. Observations have shown that the horizontal wavelength of the waves is normally about 5–15 km and most commonly around 10 km (see Figure 4.11). Such mountain or *lee waves* appear to require the following conditions. First, for strong and well-developed waves an essential requirement is marked air stability where the air is disturbed by the obstacle. Suitable conditions can be created when an isothermal layer exists (that is, no change of temperature with height) or sometimes when a temperature inversion with lesser stability above and below the stable layer exists. Second, wind direction and speed are important considerations. In the case of a long mountain ridge, the direction of flow must be more or less perpendicular to the ridge (that is, within about 30° of the perpendicular) and be maintained throughout a considerable depth in the atmosphere for strong lee waves to form. Moreover, a minimum speed of about 10 m/s at the crest level seems to be a good approximation for lee waves to begin.

In moist air, lee waves manifest themselves by cloud forming near the wave crest (Figure 4.11). Ascending and cooling air on the windward side of the wave produce semi-stationary clouds, which dissolve and dissipate as air descends and warms on the leeward side. Under such clouds, there

Figure 4.11 Lee waves and lenticular clouds: (a) cap cloud, (b) lenticular (roll) cloud, (c) altocumulus lenticularis, (d) cirrus, (e) main rotor, (m–n) wavelength. (Source: Adapted from Ernst, 1976, by permission of the American Meteorological Society)

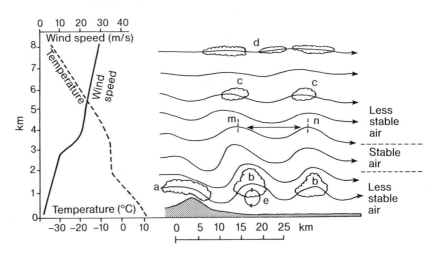

Plate 4.4 Altocumulus lenticularis developing in lee waves. Three distinct lee waves are seen in this photograph, taken looking west from Boulder, Colorado on 29 August 1982. Strong westerly winds were blowing over the continental divide of the Rocky Mountains, just west of Boulder. Each lee wave is identified by lenticular clouds forming near the wave crests, where air is ascending and cooling. Clouds dissolve and dissipate in the intervening areas, as air here descends and warms. (Photograph: Ronald L. Holle)

Figure 4.12 The föhn effect when a parcel of air is forced to cross a mountain range. (Source: Flohn, 1969, p. 47)

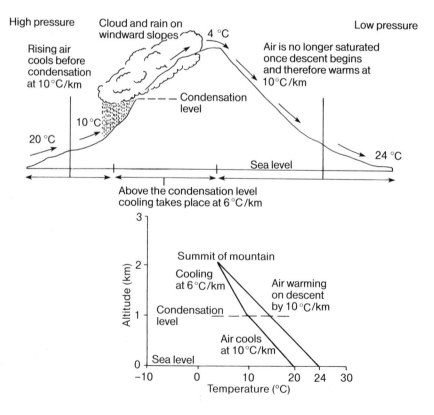

are associated overturnings of air called *rotors*. If the mountain barrier is elongated, narrow bands of *lenticular* (lens-shaped) clouds will often appear almost equidistant apart and separated by clear spaces of almost the same size. As shown in Plate 4.4, the orientation of these cloud bands is at right angles to the air stream and aligned almost parallel to the terrain barrier of the Rocky Mountains.

2. The *föhn* effect

The name *föhn* is now used as a general term for a strong, gusty, hot, dry wind which periodically occurs in the lee of a mountain range. A föhn-type wind may flow across a mountain range, if a pressure difference develops between the windward and leeward slopes. This may occur when a cyclone (region of low pressure) moves into, or develops in, the lee of a mountain range. Air is drawn towards the low pressure from the top of the ridge or beyond (see Figure 4.12).

(a) *Reasons for the föhn effect*

There are three reasons why the föhn is drier and warmer than the wind on the windward slope at a similar altitude. First, air that is forced to rise on the windward slope will cool rapidly at the dry adiabatic lapse rate (10°C/km), until it reaches its dew point and condensation occurs. Thereafter, because of the release of latent heat of condensation, the air cools at the slower saturated adiabatic lapse rate (6°C/km). Second, if moisture is then removed in the form of precipitation on the windward side, the air will no longer be saturated as it approaches the lee descent. Thus, third, during the descent on the lee slope, the air warms from the outset at the more rapid DALR.

(b) *The föhn in action*

A föhn can start very abruptly in a matter of minutes, as at Obihiro, in Japan, on 26 May 1963 (Figure 4.13) or gradually over several hours. If the start of the föhn is sudden, temperatures may rise by 20°C in a few minutes and relative humidity may drop to as low as 10%. Changes in wind speed from less than 5 m/s to over 45 m/s in a matter of seconds have been recorded. Once a föhn has set in, it tends to be very gusty, but maintains a fairly steady direction. Such conditions can cause rapid snow melt and may lead to avalanches and flooding in winter, and forest fires, soil and seed desiccation in spring.

The best-known example of this type of wind is the Föhn (from which the general name has been taken), which flows down the northern slopes (that is, the South Föhn) and sometimes the southern and western slopes (the North Föhn) of the Alps in winter and spring. Satellite imagery of northern Italy can be used to demonstrate the Alpine Föhn. As shown in Plate 4.5, when the North Föhn is blowing across the Alps from the north and west, cloud covers much of the northern (windward) Alpine slopes and summit areas (at M). The Po Valley in the lee of the mountain range is

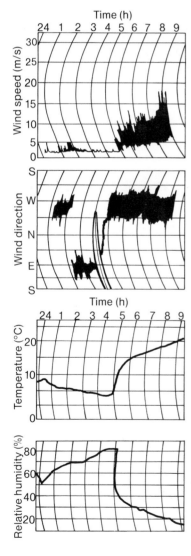

Figure 4.13 Rapid changes in wind speed and direction, temperature and humidity as a result of the abrupt start of a föhn on 26 May 1963 at Obihiro, Japan. (Source: Atkinson, 1981, p. 84)

Plate 4.5   The orographic effect of the Alps and Apennines on the development of clouds. Cloud covers much of the northern and windward Alpine slopes (M), whereas the Po valley region is cloud-free (N), due to the föhn effect. A similar effect can be identified over the Apennines (ABC), which is cloud-covered on the windward slopes, while downwind, orographic wave (lenticular) clouds, forming at right angles to the terrain barrier, are present. Black arrows indicate wind direction. (METEOSAT image supplied by the European Space Agency)

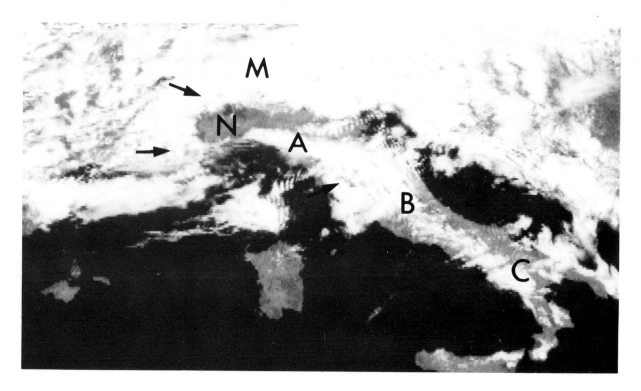

cloud-free, indicated by the clear area N on the plate. Other well-known occurrences of the föhn effect include the Chinook wind, east of the Rockies of North America, and the Zonda in the eastern foothills of the Andes.

ASSIGNMENTS
1. *Refer to Figure 4.11 and the text.*
    (a) *Describe what is meant by lee waves.*
    (b) *Analyse the terrain and meteorological conditions necessary for their formation.*
    (c) *Describe and explain the development and distribution of the cloud forms shown.*
2. *Refer to Figures 4.12 and 4.13.*
    (a) *Give a reasoned account of the föhn effect.*
    (b) *Describe the meteorological changes associated with the onset of the föhn wind at Obihiro, Japan on 26 May 1963.*
    (c) *Outline some environmental consequences of the föhn wind.*

Key Ideas

A. The Hydrological Cycle

1. The hydrological cycle is the global cyclical exchange of moisture and heat between the land, sea and the atmosphere.
2. The hydrological cycle is composed of a network of stores, in which moisture is held, and a series of moisture exchanges between and within stores.
3. The amount of moisture (mostly water vapour) in the atmospheric compartment is relatively small, but is highly varied in time and space.
4. The ability of the air to hold water vapour depends on its temperature, with warm air holding more water vapour than cold air.
5. When air is holding the maximum amount of water vapour possible, it is said to be saturated.
6. The water-vapour content of the air can be expressed in terms of its specific humidity, which is the ratio of the mass of water vapour to the total mass of moist air.
7. The relative humidity, on the other hand, is the percentage ratio between the actual specific humidity and the maximum (saturated) specific humidity.
8. Evaporation and condensation are two important phase changes in the hydrological cycle and are accompanied by the absorption and liberation respectively of latent heat.
9. Water is removed from the ocean and land surface by the turbulent transfer of evaporated moisture, condensed to form clouds and sent back to the earth's surface by precipitation.
10. The transfer and transformation of moisture between stores is considerable, but in overall balance.

B. Condensation: Basic Mechanisms

1. The main forms of condensation are clouds, mostly at high levels, and fogs, at or near the ground.
2. These forms of condensation occur when air is brought to saturation point or its dew-point temperature.
3. Most saturation and condensation in the atmosphere takes place as a result of air cooling.
4. Apart from air mixing, air can be chilled to produce ground fogs and low cloud as a result of contact cooling with the ground by advection and radiation.
5. The principal cause of air cooling, and the formation of most condensation (cloud) in the atmosphere, is the vertical ascent of air.
6. Air is forced to rise and cool by one or several of the following mechanisms: (a) orographic and frontal uplift, (b) large-scale convergence and ascent in low-pressure systems, and (c) smaller-scale convective currents.

7. Condensation is assisted in the atmosphere by condensation nuclei, around which liquid water can form.
8. Freezing nuclei also help in the formation of ice crystals.
9. Clouds and fogs are composed of any or all of the following: water droplets above 0°C, supercooled water droplets below 0°C and ice crystals well below freezing.

C. Condensation: Subsequent Development

1. The environmental lapse rate (ELR) is the actual rate of temperature change with height in the atmosphere.
2. When a vertically displaced parcel of air alters in temperature due to internal pressure change and not because of heat exchange with the surrounding air, the process is called adiabatic.
3. The dry adiabatic lapse rate (DALR) is the rate of temperature decrease (or increase) in a rising, expanding (descending, compressed) parcel of unsaturated air as it moves vertically through the atmosphere.
4. The saturated adiabatic lapse rate (SALR) is the rate of temperature decrease (or increase) in a rising, expanding (descending, compressed) parcel of saturated air as it moves vertically through the atmosphere.
5. The relationship between the ELR and both the DALR and the SALR influences the temperature, and thus the density and buoyancy of a vertically displaced parcel of air, in comparison with its atmospheric surroundings. It thus determines the character of air stability.
6. 'Unstable air' occurs when a vertically displaced air parcel is encouraged to rise still further, if initially uplifted, or if forced downward, to continue descent. In this case, the ELR is greater than the DALR and SALR.
7. With 'stable air' a parcel of air displaced vertically upwards or downwards in the atmosphere will tend to return to its original position. Here the ELR is less than the DALR and SALR.
8. Air stability is an important concept, because it determines the buoyancy of the air and thus the development of cloud and fog.
9. Cumulus clouds, often of large vertical extent, are characteristic of unstable air, whereas horizontally developed stratus clouds tend to form in more stable air.

D. Forms of Condensation

1. When air is cooled to dew point by advection and radiation under highly stable atmospheric conditions, advection and radiation fogs occur.
2. Radiation fogs are particularly found in, and intensified by, valley locations and urban industrial areas.
3. Clouds can be classified on the basis of their form, composition and altitude of occurrence.

4. Three major cloud types include stratiform, which are extensive, laterally developed clouds; cumuliform, which often occur singly and are of greater vertical extent; and cirriform, which are high-altitude clouds.

E. Orographic Effects

1. When an air stream is forced to rise over a mountain barrier, lee waves may develop.
2. The development of lee waves is determined by air stability, wind speed and direction.
3. In moist air, regularly spaced lens-shaped (lenticular) clouds form at the crests of the lee waves.
4. A strong, gusty, hot, dry wind called a föhn may form periodically on the lee slope of a mountain barrier.
5. The low relative humidity and high temperature of the föhn wind is a result of (i) slow cooling at the SALR during part of the ascent of the windward slope, (ii) the removal of moisture by precipitation at the summit and (iii) rapid warming at the DALR over the whole of the leeward descent.
6. Föhn winds may occur abruptly and are associated with rapid snow melt, and an increased risk of soil and vegetation desiccation.

Additional Activities

1. Examine the temperature sounding or vertical profile of the atmosphere shown in Table 4.1. The following are known to occur.
   - (i) Three parcels of air are forced to rise from ground level through such an atmosphere.
   - (ii) The air parcels have a surface temperature of respectively 7°C, 15°C and 20°C and a dew point of 2.5°C.
   - (iii) The DALR is 10°C/km and the SALR is 6°C/km.
   - (iv) The parcels continue to rise until they reach the same temperature as the surrounding air.
   - (a) On graph paper plot the ELR and the flight path of each parcel of air.
   - (b) Mark (i) the development and distribution of cloud, and (ii) the areas of air stability and instability.
   - (c) What significance does (i) dew-point temperature, (ii) the initial temperature of each air mass and (iii) the environmental lapse rate have on your results?

2. Consult Figures 4.12 and 4.13, Table 4.2 and Plate 4.5.
   - (a) Using a good atlas, map the location of the north Italian, Swiss and German stations in Table 4.2.

Table 4.1 Temperature sounding or profile of the atmosphere

Height (m)	Temperature (°C)
Ground level	5.0
250	3.0
500	0.5
750	0.0
1000	2.5
1250	5.0
1500	4.0
1750	0.0
2000	−1.0
2250	−2.5
2500	−3.0
2750	−5.0
3000	−2.5

Table 4.2 Mean temperature at selected stations during 12 cases each of North Föhn and South Föhn

	Lugano	Ariolo	St Gotthard	Goschenen	Altdorf	Zurich
Elevation (m)	276	1170	2096	1107	456	493
South Föhn (°C)	6.8	4.2	0.8	7.8	12.7	9.4
North Föhn (°C)	12.3	3.5	−2.5	−0.3	1.8	2.1

Temperatures adapted from potential values.
Source: Barry, 1981, p. 262

(b) Describe the meteorological changes shown in Table 4.2 associated with the onset of the Föhn wind.

(c) Use the model in Figure 4.12 to explain the observed effects.

(d) Suggest what other weather phenomena may develop in the lee of a mountain barrier, when winds are blowing directly across it.

(e) Using your results so far, (i) describe and explain the development of the clear area at N, (ii) the cloud features at M, and (iii) the cloud features at ABC.

5 Precipitation Mechanisms

A. Measurement and Dimensions

The term *precipitation* is used to denote all forms of moisture in liquid or solid form deposited directly (that is, as a precipitate) onto the surface of the earth. A distinction is sometimes made between major precipitation forms, for example rain, snow and hail, and minor forms – dew, fog-drip, frost and rime. Humans are highly dependent on precipitation for a variety of agricultural, industrial and domestic purposes, and have been preoccupied with the timing, location and magnitude of precipitation throughout historical time. It is not surprising that precipitation enhancement and control have been at the forefront of human efforts in weather modification.

1. Measurement problems

Generally speaking, only rain and snow make significant contributions to precipitation totals. Rainfall is often used interchangeably with precipitation because of the dominance of rainfall as a precipitation input, except in high latitudes and altitudes where snowfall is significant.

Precipitation remains one of the most difficult climatic elements to measure accurately, and most records must be regarded as reasonable estimates only. The rain-gauge is nothing more than a can set up to catch vertically falling raindrops and snowflakes. The 20-cm diameter, standard rain-gauge covers only a minute fraction of a square kilometre (3.3×10^{-8} km^2), and thus is strictly representative of rainfall conditions only within a very small area. In addition, a good deal of error is introduced into rain-catch totals because of the differences in rain-gauge location, size and height above ground level, wind and turbulence, rainsplash and evaporation. Snow recordings are particularly prone to sizeable error (up to 50%), because of wind and turbulence. Even with rainfall, it is not uncommon for site exposure problems to generate errors up to 20–30%.

2. Dimensions of rainfall

The measurement and occurrence of rainfall are usually described using one or more of the following physical dimensions or characteristics.

(a) Duration

This is the time period over which rain may be said to occur. Most individual rainfail events last up to several hours. They tend to be shorter where rainfall is caused by the diurnal cycle of heating and cooling (that is, convective activity) and longer in regions affected by cyclonic disturbances. Rainfall duration can be measured over longer time-spans and not just for individual rainstorms. For instance, the dryness of lowland England is well illustrated by the 407 hours per year of rain at Ipswich, in contrast to wetter areas, such as Loch Sloy in western Scotland with 1422 hours per year (see Table 5.1).

Table 5.1 Rainfall amount and duration at selected stations in the UK

Station	Altitude (m)	Mean annual duration of rainfall (h)	Mean annual rainfall (mm)
Ipswich	38	407	619
Leeds	100	552	748
Loch Sloy	12	1422	2699
Swansea	8	696	1092

Source: HMSO, 1973

(b) Depth

Rainfall is most often expressed as a depth, as if it were measured on a level surface, in millimetres. Rainfall episodes bringing less than 0.2 mm of water are called 'traces', being too small to measure in most gauges. When 0.2 mm or more of water falls within a 24-hour period, starting at 0900 hours, the period is called a rain day.

Depth can be measured for individual storms: for example, a heavy storm may deposit 100 mm or more of water. Longer periods can also be

Figure 5.1 Distribution of rainfall (mm) as a result of a large storm during 19/20 July 1977 over Johnstown, Pennsylvania: (a) as observed by using rain-gauges; (b) as estimated from infra-red imagery. (Source: Woodley *et al*. 1978)

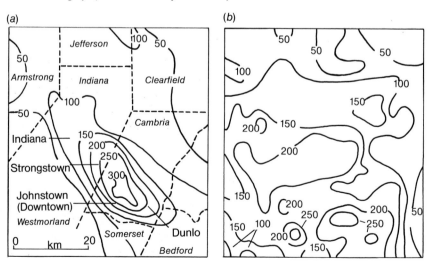

considered. For instance, the average yearly rainfall at Swansea (a relatively wet western station in the UK) is 1092 mm, whereas it is 619 mm at Ipswich, in the drier eastern part of the country.

Figure 5.1 shows the spatial distribution of rainfall depth resulting from a 10-hour storm over Johnstown, Pennsylvania, USA on 19/20 July 1977. Measurements were taken using both rain-gauges and radar. Although the two patterns of rainfall depth differ in detail (a reflection of the difficulty of measuring rainfall itself), an important point stands out. The maximum amount of water contributed by the storm (over 300 mm) tends to be located over a relatively small area. When storm rainfall is mapped over progressively larger areas from the central zone, rainfall amount falls off very rapidly.

It is also possible to average storm rainfall depth over different-sized areas and for different time periods. Figure 5.2 indicates the area–depth relations for two very large storms. It is clear from this that the average depth of rainfall from storms of different type and duration decreases very sharply as the area under consideration increases.

Figure 5.2   Area–depth relations for two very large storms. (*a*) Hurricane Camille, 19/20 August 1969 in Virginia for 17 hours; (*b*) a storm of 20–23 September 1941 in New Mexico over 24 hours and (*c*) over the total storm period of 78 hours. (Source: Barrett and Martin, 1981, p. 6)

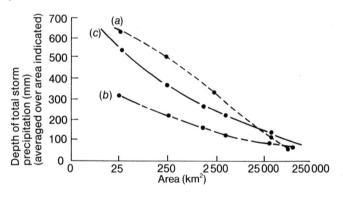

(c)   Intensity

This is the amount of rainfall in unit time (or the rate of rainfall delivery). There is a wide range of intensities of rain rates, from practically zero to above 100 mm/h. The highest intensities are recorded over the shortest time intervals. Intensity also varies according to its spatial occurrence. For instance, in the centre of the storm intensities may be high, but as a larger and larger area is considered, average intensities will decrease.

(d)   Frequency

The number of times a specific amount of rainfall (of any dimension) occurs in a given area is referred to as the frequency of precipitation. Thus the likelihood of rain falling on any day in the English Lake District, where there are more than 200 rain days per annum (that is, a high-

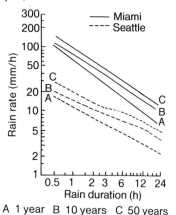

Figure 5.3 Duration–intensity curves of rain for return periods of 1, 10 and 50 years at Seattle and Miami. (Source: Barrett and Martin, 1981, p. 6)

A 1 year B 10 years C 50 years

frequency area) is far greater than rain falling on any day in a desert area which records only a few rain days per annum (low frequency). Moreover, heavy storms are more frequent, and therefore more likely, in the tropics than in temperate areas.

A quantitative relationship between the frequency of occurrence and the amount or intensity of precipitation is given in Figure 5.3. This shows, on the basis of a long series of observations, the average time period or *return period* within which a specified amount or intensity of rain can be expected to occur once, at Seattle (Washington) and Miami (Florida). The curves indicate that only once in ten years at Seattle is it expected that rainfall, averaged over a 12-hour period, will reach or exceed 5.5 mm/h. This is about one-third the intensity expected for a ten-year 12-hour storm at Miami, which is about 15 mm/h. It is important to recognise that such data refer to the average return periods. There is no reason why a ten-year 12-hour storm giving 5.5 mm or more of water at Seattle could not occur more than once in any one year or even over several days.

ASSIGNMENTS
1. *Define what is meant by the duration, depth, intensity, frequency and the return period of rainfall. Indicate how rainfall frequency is related to the concept of the return period.*
2. *Refer to Figure 5.3.*
   (a) *Describe the relationship shown between the duration–intensity rainfall curves and the average return periods.*
   (b) *Calculate the expected rainfall intensity at Seattle and Miami for a 6-hour storm: (i) occurring once per year; (ii) once every ten years; (iii) once every fifty years.*
   (c) *What do your results tell you about the mean rainfall intensity at Seattle and Miami?*
   (d) *Explain why (i) the expected rainfall intensity associated with a fifty-year storm is greater than that for a one-year storm; (ii) the expected rainfall intensity is greater for a one-hour storm than for a 12 or 14-hour storm.*

B. Formation and Forms of Precipitation

1. Theories of precipitation formation

There are four conditions necessary for the formation of major precipitation (that is, rain, snow and hail). These involve mechanisms that produce: (i) air cooling, (ii) condensation and cloud formation, (iii) an accumulation of moisture and (iv) the growth of cloud droplets. We have already seen in the previous chapter that conditions (i) and (ii) occur usually without too much difficulty. When clouds form in the atmosphere, however, they are non-precipitating in 99% of cases. Clearly stages (iii) and (iv) are fundamental in precipitation production.

With regard to stage (iii), there is not sufficient water vapour within any single cloud to sustain precipitation for any length of time. An organised thunderstorm, for instance, concentrates water vapour from 1000 km^2 around it, thereby inhibiting the growth of other cloud systems. In this way, many small clouds of late morning can be replaced by one large system in the afternoon that has starved out its neighbours. Thus, water vapour is continually being drawn into rain clouds from the surrounding environment in order to sustain them for precipitation.

Stage (iv) is perhaps the most critical one in precipitation formation. The water droplets and ice crystals of clouds have to be transformed into heavier particles if they are to fall out of clouds as precipitation. There are two main mechanisms by which cloud particles (average diameter about 0.1 mm) increase to a size suitable for precipitation (average diameter 0.5–2.0 mm). These mechanisms sometimes act singly and sometimes together.

(a) *Growth of cloud droplets*

(i) *Collision mechanisms.* The more easily understood method of particle-size increase is by *collision processes*. These mechanisms depend on the sweeping up of a mass of tiny cloud particles by a smaller number of larger particles. Rising and sinking air motions within clouds carry with them cloud particles of different sizes. Because of friction, particles of different sizes move at different speeds, and this leads to collisions between particles. The larger droplets tend to catch more of the smaller cloud particles and grow at the latter's expense. The growth of cloud particles by collision takes different paths. When two liquid water droplets collide and join together the process is called *coalescence*. The conjoining of two ice crystals is known as *aggregation*, whereas if an ice crystal collects a water droplet the process is known as *accretion*. Of the three main categories of precipitation formation, rainfall results largely from coalescence, snowfall from aggregation and hail from accretion.

(ii) *Ice-crystal method.* The second way in which cloud particles may grow to precipitation size is referred to as the *Bergeron–Findeisen, or the ice-crystal mechanism*. Although supercooled water droplets (that is, those below 0°C) and ice crystals can coexist within a cloud, they are unstable with respect to each other. What tends to happen is that liquid water droplets evaporate. This water vapour then condenses and freezes onto the surface of the moving and growing ice crystals. These ice crystals then combine by aggregation into large snowflakes. When falling to lower and warmer levels, these snowflakes may melt and continue their descent as large raindrops. If the freezing level is at the surface, or if the air throughout remains very cold, the snowflakes will not melt, producing snowfall.

In the tropics, rain frequently falls from clouds devoid of ice crystals. In these cases, raindrops grow by collision processes. Most clouds in extratropical latitudes, though, especially those in the upper, colder layers of the atmosphere, contain both ice crystals and water droplets. It has therefore been suggested that the ice-crystal method is responsible for much of the heavy rainfall of mid-latitude areas.

2. Forms of precipitation

The cloud-based precipitation mechanisms outlined above operate for the major forms of precipitation (rain, snow and hail). Rather different in character and mode of origin are the minor precipitation forms such as dew, frost, fog-drip and rime.

(a) Major categories

Water droplets normally greater than 0.5 mm – typically 1–2 mm diameter – falling to the earth's surface are defined as *rain*. *Drizzle* is a type of rain consisting of many very small particles with a diameter less than 0.5 mm; drizzle yields only traces of water.

In winter, when temperatures remain low in the atmosphere, ice crystals falling from cloud layers do not melt. They may reach the ground as *snow*, the winter counterpart of rain. Snowflakes form from the aggregation of many ice crystals and develop only in relatively calm conditions inside a cloud. They are associated with stratus, especially altostratus clouds, and less active cumulus. *Sleet* is a mixture of snow and rain, and will often reach the surface when temperatures are as high as 3–4°C. Conversely, when raindrops fall through a layer near the surface with temperatures below freezing, the airborne precipitate may freeze to form a solid sheet of ice called *glazed frost*. On a road surface, the same process produces the notorious *black ice*.

Hail consists of large, roughly spherical ice pellets, 5–50 mm or more in diameter, showing a layered structure of opaque and clear ice in cross-section. Hailstones are found when supercooled water droplets join by accretion with ice crystals within a cloud. The only cloud system which produces hail is the cumulonimbus; not all such systems give hail, however, as their main precipitate is abundant rainfall.

Such vertically extensive cloud systems are required for hail formation because they ensure that both water droplets at low levels and ice crystals in the upper layers are present within the same cloud system. The cumulonimbus system also provides the very strong updraughts of air which are required to hold the ice pellets against the force of gravity, long enough to increase to the size of hail before continuing a downward path to the earth's surface.

The extremely rapid and continuous uplift necessary for hail is found under unstable atmospheric conditions. This will occur, for instance, within a moist air mass with a very steep environmental lapse rate (see Chapter 4, section C). Such conditions are found, in particular, in mid-continental interiors, for example in the USA and the USSR between 30 °N and 60 °N. Here, in spring and early summer the water content of the cumulonimbus is already high and the temperature lapse rates are largest. Significantly, in these areas most hail occurs in the afternoon, following maximum ground heating and atmospheric instability. Countries affected by hail storms (for example, USA, USSR and Italy) have taken steps to suppress them. This is not surprising, since some hailstones can weigh up to 2.5 kg and fall at rates in excess of 100 km/h, causing severe damage to crops, property and even life.

(b) *Minor categories*

Dew is one form of precipitation that does not need visible cloud development. It forms when wind speed near the ground (about 2 m altitude) is less than 10 km/h, and when the air in contact with the ground is cooled to saturation. Dew forms on the leaves of plants, especially on cool grass blades, which act as condensation nuclei. The water vapour which condenses to form a precipitate on the surface comes partly from the air and partly by evaporation from the soil. In many semi-arid and arid areas (for example, the Atacama Desert in Chile), dew produced by nocturnal cooling can be an important source of moisture for plants and animals alike.

If dew later freezes on the ground, or if the air is very cold and dry, the water vapour condenses directly by sublimation as a fine ice deposit in a variety of crystalline forms. In either case, the precipitate is known as *hoar frost*.

When low clouds and fogs drift through trees, fog droplets are induced to precipitate against branches and leaves. Most of these droplets may later cascade to the ground as *fog-drip*. In regions such as coastal California and the Atacama Desert, where sea fog tends to be persistent, fog-drip is very important. Along the Californian coastal belt, fog-drip may account for about one-half of the annual total precipitation for all sources. Without this moisture supply, equivalent to about 250 mm, the giant redwood forests of the coastal mountains could not survive. Finally, *rime* forms when supercooled liquid fog droplets freeze when carried by wind against solid terrestrial objects such as trees and park benches.

ASSIGNMENTS
1. (a) *Outline four conditions or stages necessary for major precipitation.*
    (b) *Describe the two methods by which cloud particles are thought to grow to precipitation size.*
    (c) *How do these processes differ in low and high latitudes?*
2. (a) *Describe the character and mode of formation of the following forms of precipitation: rain, snow, hail, dew, hoar frost, fog-drip and rime.*
    (b) *Outline any broad differences and similarities.*

C. Precipitation Types

Air cooling by vertical ascent may result not only in condensation and cloud development, but also – under the circumstances outlined in the previous section – in precipitation. Accordingly, it is usual to recognise three main types of precipitation, depending on the mode of air uplift. All three mechanisms frequently act together and rarely in isolation of each other.

1. Stable upglide precipitation

This first type is variously called *convergent or cyclonic precipitation*. It is relatively steady and moderate in intensity and amount. This type is promoted, first, by widespread, dynamically forced uplift of moist air over large areas, in association with low atmospheric pressure as in depressions; and second, by air-stream convergence along frontal zones in the mid-latitudes.

(a) Convergent precipitation at the ITCZ

The horizontal convergence of air streams within a low-pressure area takes place along the intertropical convergence zone (ITCZ) (see Plates 2.1 and 2.2). In this equatorial low-pressure zone, low-level air-stream convergence in the easterlies gives rise to high-level westward-moving bands of convective precipitation with cumuliform clouds. Cyclonic or

Figure 5.4 Distribution of mean annual rainfall (mm) in the British Isles (1931–60). (Source: Atkinson and Smithson, 1976, p. 165)

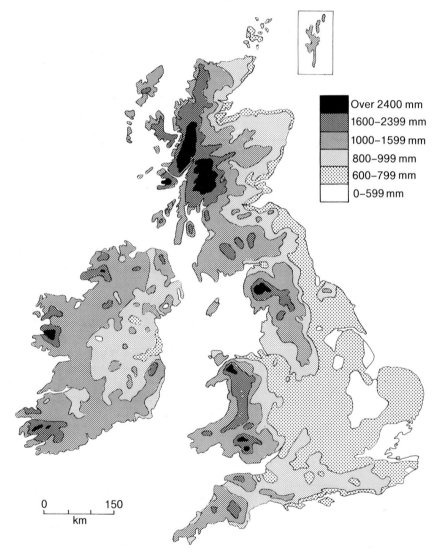

Over 2400 mm
1600–2399 mm
1000–1599 mm
800–999 mm
600–799 mm
0–599 mm

0   150
km

convergent precipitation here is greatly augmented by convective activity and is examined in greater detail below.

(b) Cyclonic and frontal precipitation

In extratropical areas, between about 40° and 65° latitude, there is an important belt of low-pressure systems, moving from west to east. Such mid-latitude depressions give rise to moderate and generally continuous precipitation over very extensive areas.

When warm air comes into contact with cold air along the air-mass boundary at the polar front, the warmer, less dense air is forced to rise over the colder, denser air (see Figure 4.5 and section A of Chapter 7). In the advancing or forward sector of the depression, at the warm front, warm air that is forced to upglide and cool over the colder air gives rise to a multi-layered cloud of the nimbostratus type. This may give continuous light-to-moderate precipitation over large areas and may last 6–12 hours or more, depending on the speed of the depression. In the rear sector, at the cold front, cold air tends to undercut the warmer air. Precipitation at the cold front is associated often with cumuliform cloud and is characterised by a shorter period of heavy showers, sometimes accompanied by thunder. The nature of this more intense precipitation is related to the steeper frontal zone and hence to the faster ascent of the warmer air.

The nature and distribution of rainfall in the British Isles is primarily explained in terms of the west–east passage of depressions with their dynamically induced air uplift over the area. Indeed, more than 70% of the precipitation of this area can be classified as cyclonic or frontal. As shown in Figure 5.4, there is a general decrease of mean annual rainfall amount from west to east and from north to south. As most depressions track from west to east, they tend to deposit most moisture in western (and higher) districts. As many of them travel in a general direction towards Norway and pass over Britain on its northern flank, higher and more prolonged rainfall occurs in northern than in southern locations.

2. Buoyancy or convective precipitation

In contrast to stable upglide rainfall, precipitation resulting from convective, buoyant updraughts of air is both localised and short-lived, and is associated with showers and heavy downpours from cumulus and cumulonimbus clouds. Convective or buoyancy precipitation is usually classified on the basis of the spatial distribution and character of the precipitation itself.

(a) Small-scale convection

Summer heating over the land surface is the main cause of local instability in the lower atmosphere, and for the development of scattered convective cells producing showers and thunderstorms. In relatively stable air (see Figure 4.10), precipitation may ensue from weak, convective clouds giving the short-lived summer showers of about 10 minutes' duration. In unstable air (see Figure 4.8), convective currents rising unchecked may produce cumulonimbus clouds, from which heavy downpours and giant hail can result. Limited areas are affected (20–50 km²) by such individual heavy storms, which may last up to one hour and yield in excess of 25 mm/h.

In Britain, buoyancy precipitation is most pronounced during the summer months in south-east England, where the warmest temperatures prevail (see Figure 2.16). Indeed, 52% of the annual precipitation of lowland Britain occurs in the five months from May to September, whereas in highland Scotland only 35% of the annual total occurs in the same period. This is because the latter area is dominated by frontal and cyclonic activity, particularly in the winter period.

South-east England has the highest frequency of short-duration (up to half-an-hour), high-intensity rainfall. Some of the very heaviest daily rainfall figures in the British Isles (from storms of about two hours' duration) occur over the south-west peninsula in Devon, Dorset and Somerset in summer. These heavy falls originate from the passage of convective thunderstorms moving in from the south. Such convective storms are augmented by two factors. First, they are fed by a long sea fetch to the south-west of Britain, from which copious supplies of moisture and energy can be picked up, and then they are intensified by mountain barrier effects, as they pass over the upland peninsula of south-west England.

(b) Large-scale convection

Small-scale convection, if persistent and vigorous, can maintain large-scale circulation patterns. At the equator, in the region of the ITCZ, strong convective activity contributes to, and is affected by, mass airstream convergence in the easterlies. Within tropical cyclones along the ITCZ, especially over tropical oceans, convective cells produce cumulonimbus clouds, each individual cloud having a diameter of up to 10 km. These convective cells may group into larger structures: first, into convective units of 100 km across and eventually into cloud clusters 100–1000 km in diameter. Rainfall from these convergent–convective cloud clusters can be very heavy and prolonged, affecting areas of thousands of square kilometres.

3. Orographic precipitation

There is little doubt that cyclonic, frontal and convective activities are greatly altered and augmented by relief barriers. So much so that a separate 'type' of precipitation has been identified: orographic rainfall. There are three reasons why rainfall values are generally higher over upland than lowland areas. First, relief may prolong depression activity in an area by retarding the rate of movement of a depression. The longer a depression remains in an area, the greater will be that area's rainfall. Second, mountains act as a barrier to moist air streams, which are therefore forced to rise and cool over them. Third, hills act, especially in the summer, as high-level heat sources. On sunny days, convective cells tend to develop over them, giving showers and thunderstorms of the kind outlined above.

(a) Orographic precipitation model

A well-known, but simple meteorological model implies that when mountain barriers intercept rain-bearing systems, rainfall increases with altitude, especially on the windward side, and marked reductions occur in the *rain shadow* of the lee slope (Figure 5.5). In the middle and high latitudes, this effect is particularly pronounced on west-coast mountain ranges which lie in the path of west to east-moving frontal depressions. In the Olympic Mountains of Washington State, USA, precipitation increases until the air stream has passed over the summits (2300–2700 m), and thereafter precipitation decreases as the air stream descends onto the leeward side (see Figure 6.7). The increase of precipitation with altitude in Britain has long been recognised. It is an effect which dominates precipitation distribution, not only on annual maps (Figure 5.4), but also on monthly and many daily precipitation maps as well.

Figure 5.5 An orographic precipitation model. An increase of rainfall with height is shown on the windward slope, and a reduction in rainfall amount occurs in the rain shadow of the lee slope (see also Figure 6.7).

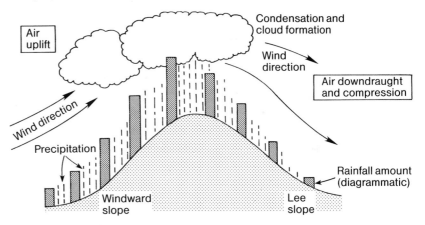

(b) The model revised

The relationship between relief and precipitation is highly complicated. The above model needs to be replaced by more realistic models, for a variety of atmospheric conditions, in each particular mountain area. The main adjustments include the following.

1. Over narrow uplands (for example, a transverse ridge), the horizontal scale may be insufficient to trigger maximum cloud and precipitation build-up.
2. Precipitation may be carried over the crest line by the wind, causing a leeside maximum. In western Britain, with mountains of about 1000 m the maximum falls are recorded to the leeward of summits.

109

3. In mid-latitudes, the zone of maximum precipitation on west-coast mountains may occur at about 2000–2500 m (that is, the summit of the Olympic range), but will decrease thereafter on higher mountain summits. This is possibly the case on the high summit slopes of the mountains of British Columbia and in the Sierra Nevada of California.

4. Further inland from the west-coast mountain ranges, the maximum precipitation may occur significantly below the summit levels. This may apply in the Rockies, for instance, where winds from the west, after shedding their moisture and becoming dry over the coastal ranges, are unable to sustain maximum precipitation inland on the upper slopes of the Rockies.

5. In the tropics and subtropics, in particular, the maximum precipitation often takes place well below the highest peaks (see Figure 5.6).

Figure 5.6 Generalised profile of mean annual precipitation with height in a number of mountain areas in the tropics. (Source: Barry, 1981, p. 186)

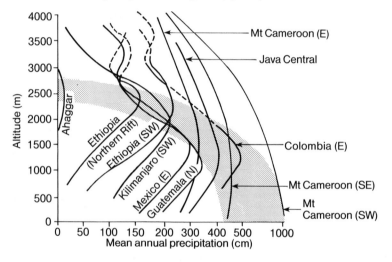

In Colombia and Guatemala, there is a marked decrease at about 1000–1250 m, and at higher levels of 2000–2500 m in East Africa.

In the tropics, clouds (cumuliform types) producing orographic rain seem to concentrate and shed most of their moisture in the lower layers of the atmosphere. This is in contrast to conditions in temperate areas, where a good deal of orographic rainfall originates from stratiform clouds. These clouds (stratus, altostratus and cirrostratus) tend to release their moisture over a greater depth in the atmosphere.

ASSIGNMENTS

1. *Compare and contrast the three main types of precipitation with respect to: (i) mode of formation; (ii) associated cloud types; (iii) rainfall duration; (iv) rainfall amount and intensity; (v) spatial location and geographical extent of rainfall.*

2. (a) *Describe the model of orographic precipitation shown in Figure 5.5.*

   (b) *To what extent is this a realistic model of actual precipitation events?*

   (c) *How can the model be improved?*

3. (a) *Describe the pattern of rainfall distribution indicated in Table 5.2.*

   (b) *What are the reasons for the high frequency of heavy winter rainfall in the highland areas?*

   (c) *Account for the greater frequency of heavy summer rainfall in lowland districts.*

Table 5.2 Occasions during the period 1863–1960 with exceptionally high amounts of rain (more than 125 mm) in the rainfall day

Area	Summer half-year	Winter half-year
Dartmoor (highland)	2	2
South Wales (highland)	3	8
Snowdonia (highland)	10	15
Lake District (highland)	8	28
Western Highlands (highland)	4	23
South-west peninsula excluding Dartmoor (upland)	8	0
Northern Ireland (lowland and upland)	2	1
Other areas (mostly lowland)	26	2

Groupings according to district and season.
Source: Chandler and Gregory, 1976, p. 147

D. Interference in the Precipitation Supply

Humans can interfere with precipitation, as with the energy balance, by conscious and unconscious action. A direct conscious attempt to alter the precipitation supply is by cloud-seeding. More indirectly, humans unconsciously interfere with rainfall patterns by building cities.

1. Conscious action: cloud-seeding

(a) Precipitation enhancement

A widely known, if not yet fully accepted technique of precipitation enhancement is cloud-seeding. *Cloud-seeding* involves the introduction into clouds of condensation and freezing nuclei in order to augment precipitation. Artificial condensation nuclei such as salt particles or a fine hygroscopic mist may be added to stimulate the coalescence of small water droplets in clouds. Alternatively, freezing nuclei such as dry ice (which are frozen CO_2 pellets) and silver iodide smoke may be injected to induce freezing and the triggering of the ice-crystal process.

Although cloud-seeding experiments are being carried out all over the world, the effectiveness of their results is far from clear. While modest increases of 10–25% in precipitation have been widely reported, some experiments have resulted in a decrease of rain as a consequence of attempts to increase it. There is no way of telling, of course, if rain does follow seeding, whether any of it was in fact produced by the rain inducement. It seems that the correct amount of seeding material needs to be supplied under favourable meteorological conditions for successful results. Convective and especially orographic situations offer the highest chances of success, where clouds capable of producing precipitation are already naturally present. Little can be accomplished when skies are clear and the air is relatively dry.

(b) Hail reduction

Cloud-seeding is also being used on a growing scale to suppress hail damage. One simple, but appealing theory implies that the addition of freezing nuclei (for example, silver iodide) to storm clouds can alter the particle-size distribution, though not the total volume, of hailstones. It is thought that the additional freezing nuclei compete with the small number of natural freezing nuclei for the same finite supply of supercooled water. The result is that there should be more, but smaller – and therefore less damaging – hailstones. Despite a good deal of uncertainty surrounding the ability of cloud-seeding to reduce hail, field tests are being carried out in many countries including the USA, Italy and Australia; particular success in this area of research has been claimed by the USSR.

(c) Other uses of cloud-seeding

Cloud-seeding has been used quite successfully in fog dissipation, but unsuccessfully, as yet, in reducing the force of tropical cyclones and suppressing lightning. Cold fog dissipation is a well-established procedure. One effective method uses the heat from a jet engine to raise the temperature of foggy air along airport runways to above the dew point. Cold fogs can also be dispersed by spraying them with dry ice, or silver iodide, dropped from planes, or by liquid propane ejected from ground apparatus. By such seeding, some of the supercooled fog droplets change to ice crystals, which then grow at the expense of the water droplets. Eventually the ice crystals fall to the ground and the fog clears.

2. Unconscious action: urban effects

(a) Urban mechanisms

Humans interfere inadvertently with precipitation by building cities. A number of studies have indicated that urban industrial areas can increase local annual precipitation totals by about 10% (see Tables 2.4 and 5.3). There are four main factors that make an increase of precipitation by urban areas likely. These factors operate together and their effects are difficult to assess separately.

Table 5.3   Urban–rural differences of annual precipitation

| Locality | Precipitation | | |
	Urban (mm)	Rural (mm)	Difference (%)
Moscow, USSR	605	539	+11
Urbana, Illinois, USA	948	873	+ 9
Munich, West Germany	906	843	+ 8
Chicago, Illinois, USA	871	812	+ 7
St Louis, Missouri, USA	876	833	+ 5

Source: Landsberg, 1981

(i)   *The heat island effect.*   This is the most obvious and perhaps the most important factor. The fact that cities are warmer than their rural surroundings leads to rising vertical air motion over cities. Vertical ascent, of course, is essential for the formation of the major categories of precipitation.

(ii)   *The obstacle effect.*   A second cause of enhanced urban precipitation is the obstacle effect. The aerodynamic 'roughness' of the physical structure of urban areas can, as with relief, impede the progress of weather systems. Thus, if rain-producing mechanisms are taking place (for example, by frontal processes), they may operate longer over an urban area, giving more prolonged precipitation than in a rural spot, where the frontal systems move faster. In addition, wind speeds are lowered within cities, compared with those at the same height in the country, because of the frictional resistance offered by the urban structure. This local slowing of air causes it to 'pile up' (converge) over the city, a process which is relieved by vertical upward motion.

(iii)   *The pollution effect.*   Increased condensation and freezing nuclei are found in city atmospheres from urban pollution (see Table 2.4). If the correct quantity and sort of industrial nuclei are added to the city atmosphere, they may well contribute to enhanced cloud formation and precipitation production.

(iv)   *The absolute humidity effect.*   There is some evidence to show that absolute or specific humidities are higher in cities, especially at night, than in adjoining rural areas. The source of this moisture may be from water released by the combustion process. If this is so, such moisture may feed the development of rainstorms.

(b)   *Case-studies*

(i)   *Prevailing wind direction.*   If urban precipitation enhancement does occur, it will take time for the rainfall-producing particles to be carried up to the cloud level, and for the individual droplets to form and grow to sufficient size to fall as precipitation. Therefore, any effects are likely to occur not within, but downwind of the city itself. This theory is given support in Figure 5.7, which shows the distribution of summer rainfall totals in the St Louis area. It can be seen that maximum precipitation was reached about 20 km east of the urban district, where values were about

113

Figure 5.7 Mean summer rainfall departures in the St Louis area from that in a nearby undisturbed control site. Figures expressed in percentages. (Source: Landsberg, 1974, p. 753)

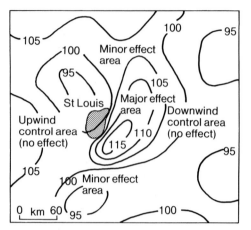

15% greater than the average for the central area of the city. Background precipitation totals of plus or minus 5% of the city centre average were found elsewhere around the city. This arrangement is in accord with the prevailing wind direction, since 90% of all rain systems move in a west—east direction at this location.

(ii) *Vertical updraught.* The heat island and mechanical updraught effects of the city are important in enhancing summer convective rainfall. In particular, they are thought to increase the frequency and intensity of severe weather such as thunder and hailstones (see Table 5.4). Occasionally, thermal and mechanical updraughts over the city can affect potentially unstable, moist air masses and frontal zones. The result is city-centred heavy shower activity. Using London as an example, Atkinson (1977) has demonstrated that the thermal effect of the city plays a major role in enhancing convective precipitation. His careful analysis makes it clear that, on 14 August 1975, general mass air convergence took place in the London area. By 1200 hours, an intense heat island had developed over the city (Figure 5.8). Screen temperatures reached 31°C within the city and were about 2.1 °C higher than anywhere else in south-east Eng-

Table 5.4 Maximum urban—rural differences in summer rainfall and severe weather events

City	Rainfall	Thunderstorms	Hailstorms
St Louis	+15	+25	+276
Chicago	+17	+42	+246
Cleveland	+27	+38	+ 90
Indianapolis	0	0	0
Washington, DC	+ 9	+36	+ 67
Houston	+ 9	+10	+430
New Orleans	+10	+27	+350
Tulsa	0	0	0
Detroit	+25	no data	no data

Values expressed as a percentage of rural figures.
Source: Landsberg, 1981

land. These temperatures were sufficient, together with unusually high atmospheric humidity, to encourage very steep lapse rates, which, in turn, promoted the development of rising air currents and deep storm clouds over the city. Shortly after 1600 hours, some of the 200 recording gauges in the city area caught the first rainfall. The heaviest downpours occurred between 1730 and 1800 hours. Figure 5.9 shows the precipitation for the 24-hour period from 0900 on 14 August; isohyets are used to join the places that have equal amounts of rainfall. The isohyetal pattern shows a very high urban maximum, reaching 169 mm at Hampstead in the northern part of the urban area. Many locations had no rain at all and, more generally, only 5 mm fell in south-east England.

Figure 5.8   Maximum screen temperature (°C) on 14 August 1975 over London and south-east England. (Source: Atkinson, 1977)

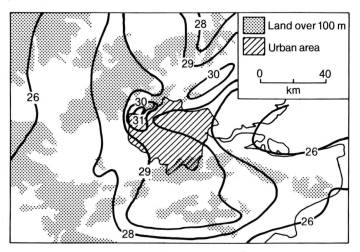

Figure 5.9   Daily precipitation (mm), from 0900 GMT on 14 August 1975 to 0900 GMT on 15 August 1975, over London and south-east England. (Source: Atkinson, 1977)

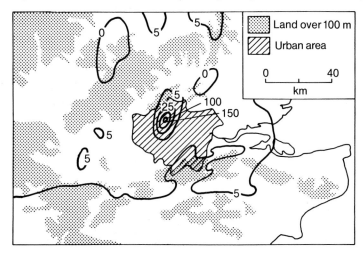

1. (a) *Examine the various ways cloud-seeding is used to modify pre-cipitation patterns.*
   (b) *List the main features that may contribute to enhanced urban precipitation.*
   (c) *What evidence is there to suggest that the urban heat island effect is particularly important in enhancing urban precipitation?*
   (d) *In what other ways might it be possible for humans to modify the precipitation regime?*

Key Ideas

A. *Measurement and Dimensions*

1. Precipitation is the term used to denote all forms of moisture de-posited from the atmosphere (i.e. as a precipitate) directly onto the earth's surface.
2. It includes major forms such as rain, snow and hail, and minor forms such as dew, frost, fog-drip and rime.
3. Precipitation is very difficult to measure accurately, and precipitation records are only estimates.
4. Rainfall is commonly expressed with reference to one or more of four basic physical dimensions: duration, amount (depth), intensity and frequency.
5. Duration is the time period during which rain falls; amount is ex-pressed as a depth in millimetres; and intensity is the amount per unit time.
6. Frequency is the number of times rainfall of any description is said to occur, and is a measure of the likelihood of rain falling in an area.

B. *Formation and Forms of Precipitation*

1. There are four conditions necessary for major precipitation, namely (i) air cooling, (ii) condensation, (iii) moisture input to cloud, and (iv) the growth of cloud droplets.
2. The critical stage is the growth of the cloud droplets, because the majority of clouds do not produce precipitation.
3. Cloud particles, comprising water droplets and ice crystals, increase to a size necessary for precipitation by two main mechanisms.
4. Water droplets and ice crystals grow by colliding with each other.
5. Cloud particles also grow to precipitation size by the ice-crystal method, by which ice particles grow at the expense of water droplets within the cloud.
6. Precipitation occurs in a variety of forms of differing appearance and mode of origin.
7. The major forms of rain, snow and hail are deposited on the earth's surface by falling from a cloud under the force of gravity.
8. The minor forms, including dew, frost, fog-drip and rime, are de-posited directly onto the earth's surface by cooling air and fog.

C. *Precipitation Types*

1. There are three principal types of precipitation, depending on the nature of air uplift.
2. Stable upglide precipitation is associated with gradual air ascent in convergent/cyclonic systems and zones of frontal activity.
3. This results in steady prolonged precipitation of moderate intensity from stratiform clouds.
4. Buoyancy or convective precipitation results largely from thermal updraughts.
5. This type of precipitation is short term and is associated with showers and heavy downpours from cumulus and cumulonimbus clouds.
6. Surface heating in the mid-latitudes produces localised convective systems, but at the ITCZ, convective and convergent processes combine to give large-scale motions.
7. Cyclonic and convective precipitation are altered by relief barriers to give orographic precipitation.
8. A simple model suggests that when rain-bearing systems are intercepted by mountain barriers, precipitation increases with altitude on the windward side with reduction to the lee.
9. There are many exceptions to this general model, especially in tropical environments, where orographic rainfall does not always increase indefinitely with height.

D. *Interference in the Precipitation Supply*

1. Humans consciously attempt to modify precipitation amounts by the process of cloud-seeding.
2. Cloud-seeding can be used to enhance precipitation, to suppress hailstone formation, and to dissipate fog.
3. The ability of cloud-seeding to enhance precipitation and to reduce hail remains unclear.
4. Humans unconsciously interfere with precipitation patterns by building urban areas.
5. Cities probably augment precipitation, especially rainfall, in their vicinity by the thermal and mechanical uplift of air, and by adding moisture and industrial nuclei to the atmosphere.
6. Precipitation enhancement usually occurs not within, but downwind of the city, because of time-lag and wind effects.
7. The thermal and mechanical effects of urban areas are thought to be important in augmenting summer convective thunderstorms and hailstorms.

Additional Activities

1. (a) Referring to Chapter 5, provide a reasoned account of precipitation in Britain with particular reference to its distribution, timing and origin.
   (b) Make a list of any aspects of precipitation in Britain not covered in this chapter.

Figure 5.10 Percentage increases in thunderstorm incidence as related to urban population size (1970). (Source: Landsberg, 1981)

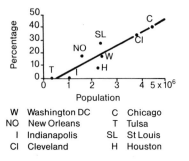

W   Washington DC   C   Chicago
NO New Orleans   T   Tulsa
I   Indianapolis   SL St Louis
CI Cleveland   H   Houston

2. (a) Describe the relationships shown in Table 5.3 and Figures 5.10 and 5.11.

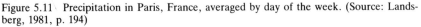

Figure 5.11 Precipitation in Paris, France, averaged by day of the week. (Source: Landsberg, 1981, p. 194)

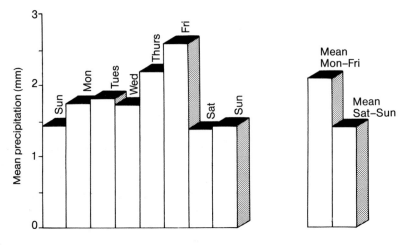

(b) Outline the evidence which supports the idea of an anthropogenic and urban origin of enhanced precipitation.

(c) Describe and explain the precipitation patterns shown in Figure 5.7.

(d) What components of urban rainfall are displayed in Table 5.4 and Figures 5.8 and 5.9?

(e) Explain the patterns indicated and show how they may further your knowledge of urban modified precipitation.

(f) With reference to Chapters 4 and 5, and other figures in this book (for example, Figure 2.14), give an explanatory account of the role of the following in urban precipitation enhancement: (i) atmospheric nuclei; (ii) moisture release from combustion; (iii) orographic effects; (iv) convective activity.

Precipitation Patterns

Introduction

Precipitation, whether expressed in terms of its duration, depth, intensity or frequency, is highly irregular both in space and time. In other words, precipitation patterns are very variable across the face of the planet. Traditionally, geographers have tended to study the spatial dimension of precipitation variability over fairly long time periods. They have examined, for instance, the distribution of mean annual and seasonal precipitation, annual variability (that is, yearly variation from the mean) and year-to-year fluctuations and trends, such as long-term drought. Studies are increasingly being made of rainfall variation over shorter time-scales, however. Short-term aspects of rainfall variation include, for instance, the exact timing and distribution of rainfall during the growing season, and the characteristics of individual rainstorms (their intensity, duration and frequency). Such short-term aspects have great practical value for farmers and hydrologists. They determine, for example, the amount of water available for soil percolation and storage, and for runoff and erosion at any particular time.

A. World Annual Rainfall Distribution

1. Zonal model

A simple model of average annual precipitation by latitude is shown in Figure 6.1. This model suggests that there is an abundance of rain in the equatorial zone, moderate to large amounts in the mid-latitude belts, and relatively low rainfall totals in the subtropics and particularly at the poles. Such zonal or north–south patterns in annual precipitation can also be identified when we turn to the more complicated world distribution map of mean annual precipitation shown in Figure 6.2.

The zonal features indicated in the world pattern of precipitation are closely linked to the model of the general circulation of the atmosphere (Chapter 3, section D). It can be seen from Figure 6.3 that rainfall is highly abundant where air uplift is encouraged. Uplift occurs in low-pressure regions in the great convergence/convection zone of the equatorial trough (ITCZ) and in the polar frontal zones of the mid-latitudes. Rainfall is scarce in areas of surface air divergence such as in the region of

Figure 6.1 Average annual precipitation (mm) by latitude for the world as a whole and for stations along the west and east coasts respectively of the American continents. (Source: Petterssen, 1969, p. 263)

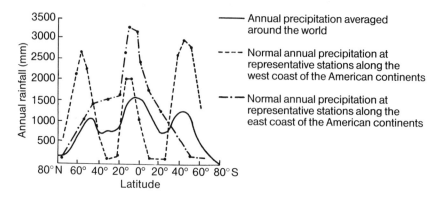

Figure 6.2 World distribution of mean annual precipitation (mm). (Source: Riehl, 1978, p. 271)

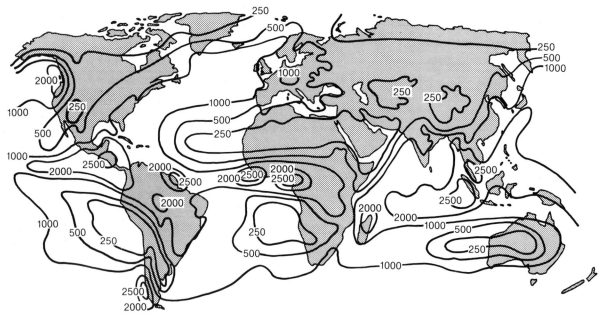

the trade-wind anticyclones and the anticyclones which dominate higher polar latitudes.

(a) *Seasonal components*

The latitudinal model of rainfall distribution is somewhat complicated by a marked seasonal migration of the equatorial trough. The equatorial trough (ITCZ) tends to follow the sun in its annual migration between the Tropic of Cancer (sun overhead on 21 June) and the Tropic of Capricorn (sun overhead on 21 December), as shown in Figure 6.4. As the whole general circulation moves in sympathy with the ITCZ, the various

Figure 6.3 Idealised longitudinal cross-section through the atmosphere depicting (*a* and *b*) the main zones of ascending and descending air motions during the seasonal extremes of winter and summer and (*c*) the associated principal areas of precipitation. (Source: Petterssen, 1969, p. 262)

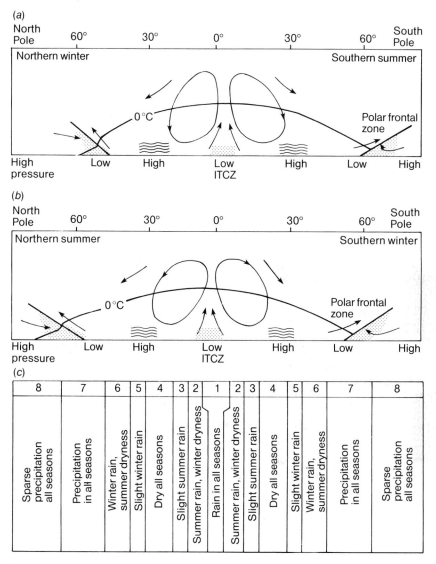

Figure 6.4 Average position of the equatorial trough or intertropical convergence zone (ITCZ) in January and July. (Source: Flohn, 1969, p. 112)

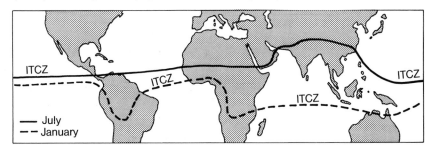

zones of vertical air uplift (wet areas) and downdraught (dry areas) migrate toward the hemisphere where the summer prevails. Plates 2.1 and 2.2 demonstrate the intertropical migration of the equatorial trough between July and December. In the former case, the ITCZ with its rain-bearing cloud systems lies about 18° N, whereas in the latter example it lies at about 18° S.

(b) Zonal model in action

From the idealised model shown in Figure 6.3, the following precipitation zones are apparent.

Zone 1. The *equatorial zone* has abundant rain throughout the year, associated with the permanence of the ITCZ. There are often two peak periods in rainfall occurrence related to the passage (twice) of the sun overhead during the year.

Zone 2. Further away from the equator, there are two zones (north and south) which receive much rain in summer and scanty rainfall in the winter. These latitudes are influenced by the ITCZ with its rain in the high sun period (summer), and by the subtropical anticyclones in the low sun period (winter). This zone is referred to as the *wet and dry tropics*.

Zone 3. Further away still from the equator, there are two belts which receive a small amount of rain in summer, but very little at other seasons. These are the *tropical semi-arid regions* on the equatorial margin of the subtropical high-pressure system. This region corresponds to the area known in Africa as the Sahel.

Zone 4. Poleward of the semi-arid belts are the *arid zones*, associated with year-long dominance of the subtropical highs and permanent dryness.

Zone 5. Poleward of the arid zones, a narrow belt of scanty winter rainfall exists. This zone is dominated by the *subtropical highs* for most of the year, but occasionally is affected by the mid-latitude depressions for a short period in winter.

Zone 6. The *Mediterranean precipitation zone* is a semi-arid or subhumid region on the poleward margin of the subtropical highs. It is characterised by a long dry summer period, when the region is affected by the subtropical highs. A short wet season prevails in the low sun period (winter), since the zone is dominated at this time by the belt of mid-latitude depressions.

Zone 7. In the *middle and high latitudes*, there is a belt where travelling depressions and fronts are dominant. Precipitation is plentiful in all seasons, but maximum precipitation occurs in the low sun period (winter), when cyclonic activity is strongest.

Zone 8. This is a *polar region* of low precipitation, dominated by cold subsiding air. Travelling depressions may penetrate the high-pressure systems in winter giving some precipitation, mostly snowfall.

2. Modifications to the zonal model

An examination of Figure 6.2 makes it quite evident that significant longitudinal or east–west variations in rainfall are superimposed on the zonal arrangement. There is a tendency, for instance, for the subtropical west coasts of continents (for example, the western Sahara) to be dry, whereas rain is more plentiful along east coasts in the same latitudes (for example, south-east Asia). (See also Figure 6.1 for the American coasts.) On the other hand, in the middle and high latitudes, the west coasts (UK, British Columbia) are generally wetter than the east coasts (Patagonia, north China, eastern USSR). Furthermore, rain is more abundant on the windward sides of mountain ranges, for example, the Rockies and Andes of America and the Western Ghats in India. Rainfall is also deficient in the centre of continental land masses that are remote from the sea in fairly high latitudes, for example, central Asia, north China. Finally, cold currents intensify aridity on land masses adjacent to them, for example, the Atacama Desert and the Namibian Desert. Some of these east–west patterns in rainfall distribution are now examined a little further.

(a) Land and ocean effects: the monsoon

As explained in Chapters 2 and 3, differences in heating and cooling between land and sea produce major modifications in the zonal arrangement of pressure belts associated with the general circulation. In the subtropics, over the eastern parts of the oceans and adjacent land masses marked subsidence of the air promotes a permanence in the subtropical highs, and thus the maintenance of arid and semi-arid conditions. In contrast, over the western sectors of the oceans and on adjacent land, the permanence of the subtropical highs is greatly weakened. In these areas, low-pressure systems tend to develop in summer. As a result, onshore winds advect moisture over the eastern parts of the continents and some quite high rainfall totals prevail.

Such monsoonal conditions are best demonstrated over the great land mass of Asia. Here, a strong winter anticyclone, associated with very cold temperatures at that time of year, gives way in summer to low pressure associated with continental heating (see Figure 3.11 (a) and (b)). Accordingly, during the northern winter, when, over Asia, the ITCZ shifts to about 21° S (see Figure 6.4), cool dry air from the interior of the continent streams southward and rain is sparse over the land (Figure 6.5). After air has moved some distance over the warm tropical and equatorial waters, showery rain becomes frequent. Thus, although India as a whole has winter dryness, rain is plentiful in the East Indies and in northern Australia (see Figure 6.5).

During the northern summer, the ITCZ shifts to about 18°N (Figure 6.4). Moist monsoon winds, originating from the southern Indian Ocean and attracted by low-pressure systems over the land, sweep north-east and then north-west across the Indian subcontinent. Rainfall is consequently heavy over India and south-east Asia as far north as the southern coasts of Japan (see Figure 6.6).

Figure 6.5 World distribution of average precipitation (mm) during the northern winter (1 November–30 April). (Source: Chandler, 1981, p. 60)

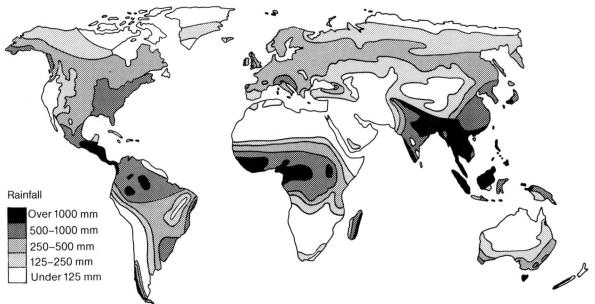

Figure 6.6 World distribution of average precipitation (mm) during the northern summer (1 May–31 October). (Source: Chandler, 1981, p. 59)

(b)   Mid-latitude cyclonic belt

In the north Atlantic, mid-latitude depressions moving west to east, charged with moisture from evaporation over a warm ocean, give a region of high mean annual precipitation extending from the Gulf of Mexico north-eastward to the British Isles, Iceland and Norway. A similar zone of high ocean temperature and high mean annual precipitation is found in the Pacific Ocean, from the Philippines towards the Gulf of Alaska. The mid-latitude depressions quickly lose their capacity to supply rain, however, as they move inland. This is because they deposit most of their moisture on the western flanks of both continents. There is, as a consequence, a general decrease in precipitation from west to east, and the interiors of the respective land masses remain relatively dry. Some greater penetration of rain occurs in Europe, when compared with North America, because of the absence of large transverse (that is, north–south) mountains acting as barriers to the rain-bearing systems. In the mid-latitudes of the southern hemisphere, where there is a general absence of major continental land masses, the region of enhanced precipitation is aligned more zonally than in the northern hemisphere.

(c) Orographic barriers

The role of mountain ranges in precipitation modification has been outlined above (pages 109–10). In the zone of the mid-latitude depressions, the western Cordillera of North America induce heavy orographic precipitation along their windward sides, north of about 35° N. At the same time, the mountain barrier depresses rainfall in an immense *rain shadow* on the leeward margins. Figure 6.7 shows that the windward slopes of the Olympic Mountains, exposed to westerly cyclones, experience about 3750 mm of precipitation per year and over 4500 mm is not uncommon. Yet Seattle, lying between the Olympic Mountains on the west and the Cascade Mountains on the east, averages only about 750 mm. As shown, this rain-shadow effect extends and intensifies further east. The lee side of the Cascades is semi-arid, and the area of the lower Columbia river, 160 km to the south, is virtually a desert. Similar orographic influences on rainfall patterns can be discerned from Figures 6.2, 6.5 and 6.6 in the region of the Andes in South America, along the Western Ghats in India and over the mountains of Madagascar.

Figure 6.7 Orographic effect on precipitation in the Seattle area. (Source: Riehl, 1978, p. 275)

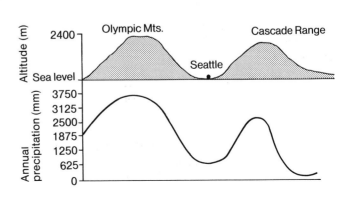

(d) Ocean currents

Although warm currents such as the Gulf Stream and Kiro Siwo may increase precipitation in the Atlantic and Pacific rain cells, the subtropical deserts are reinforced in the vicinity of cold ocean currents. The most well-known example of this effect is the Atacama Desert, along the west coast of South America in northern Chile and Peru. As shown in Figure 6.8, warm, moist air passing from the high-pressure cell over the central Pacific (Figure 3.11) towards the west coast of South America is cooled to dew point from below, over the cold water current. This surface advection cooling causes a temperature inversion to develop in the coastal regions. The resulting air stability induces a persistent fog regime below the inversion and generally prevents convective cells developing, which would create some rainfall. On moving inland over the warm land surface, the low-moisture content of the cold coastal fogs is quickly dissipated. Although the rainfall of the Atacama is almost nil, the ground does receive some moisture from the coastal region. This is in the form of dew and results when air over the Atacama Desert is chilled to saturation by nocturnal radiation cooling.

Figure 6.8 Influence of the cold Peruvian current on local precipitation in the Atacama Desert. (Source: Flohn, 1969, p. 175)

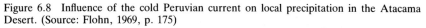

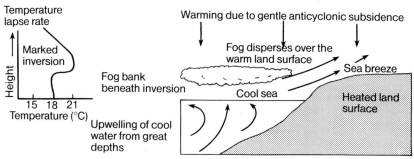

ASSIGNMENTS

1. *Examine the rainfall data shown in Figure 6.9.*
    (a) *Plot each of the listed stations on a world map, using a good atlas.*
    (b) *Using Figure 6.9, describe the annual and seasonal distribution of rainfall for each of the stations shown.*
    (c) *To what extent do your answers match the rainfall patterns shown in Figures 6.2, 6.5, 6.6?*
    (d) *For each of the stations indicated, explain the observed patterns in rainfall distribution. Make reference to (i) the model of the general circulation and its north–south oscillation, and (ii) the effects of land and sea in your answers.*

2. (a) *Using a good relief map of the world, identify from Figures 6.2, 6.5 and 6.6 those areas where precipitation is modified (both enhanced and depressed) by orographic barriers.*
    (b) *Explain the existence of the Atacama Desert with reference to (i) the permanence of the subtropical high, (ii) orographic and föhn effects and (iii) ocean currents.*

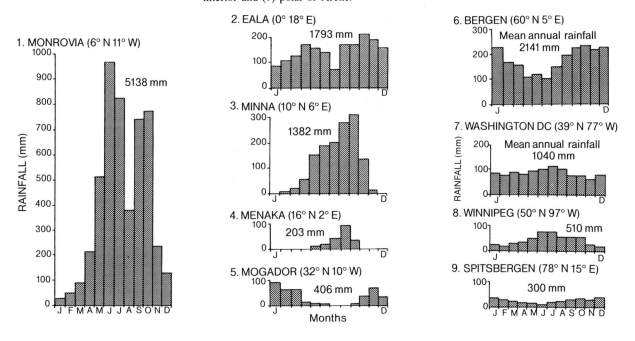

Figure 6.9 Average annual and monthly rainfall totals for selected lowland stations. Such stations are representative of the major precipitation regimes of the world: (1) tropical monsoon, (2) equatorial, (3) tropical wet and dry, (4) semi-arid and arid, (5) Mediterranean, (6) temperate oceanic west coast, (7) mid-latitude east coast, (8) temperate continental interior and (9) polar or Arctic.

B. Rainfall Variability

It is useful to bear in mind that the long-term mean annual, seasonal and monthly rainfall totals outlined in section A above give little information on the variability of rainfall supply over time. Rainfall variability is an important concept, because it gives clues to the regularity with which rain may be expected at any particular place. There are a number of ways in which rainfall variability can be expressed. If annual or monthly rainfall totals are known for a twenty or thirty-year period, these values can be depicted by the use of frequency distribution diagrams. Such values can also be plotted as a time series or trend. In studies of rainfall variability over time, it is often the characteristics of individual rainstorms which are of interest.

1. Frequency distribution diagrams

One common method of expressing rainfall variability is by the use of frequency distribution or dispersion diagrams. These diagrams depict for any given length of time the scatter of daily, monthly or yearly rainfall totals at a station by placing a dot on the vertical axis (see Figure 6.10 (a)). Alternatively, actual rainfall amounts can be grouped into class-intervals, and their distribution shown as in Figure 6.10 (b).

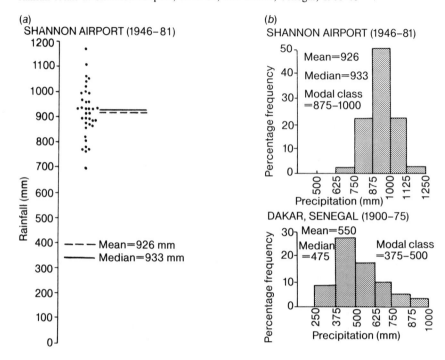

Figure 6.10 Frequency distribution diagrams depicting the scatter of (*a*) actual annual rainfall amounts at Shannon Airport, western Ireland, 1946–81 and (*b*) class-intervals of annual rainfall totals at Shannon Airport, 1946–81, and Dakar, Senegal, 1900–75

(*a*) *Annual variability*

The dispersion plots shown in Figure 6.10(*b*) indicate the percentage frequency distribution of annual precipitation at Shannon Airport, in western Ireland, and at Dakar, in Senegal. Annual rainfall distribution is much more variable at Dakar, located in a semi-arid area, than at Shannon, which is situated in a moist, temperate, oceanic rainfall regime. At Shannon, there is a close and balanced scatter of rainfall recordings about the mean value. Approximately as many values are above as below the mean value of 926 mm. Indeed, the mean or average, the median or middle value, and the mode or most frequent class all fall together within the same interval. This bell-shaped type of distribution is known as a *normal distribution*.

At Dakar, on the other hand, rainfall distribution is more irregular or asymmetrical about the mean, which lies one class-interval above the mode. Such a *skew distribution* suggests the occurrence of a few very high annual values and a large number of very low recordings.

(*b*) *Mean or median*

Because mean values can be very misleading, especially in semi-arid areas of marginal rainfall, the *median* or middle value has often been recommended. This, as suggested, is the midpoint in the series of values, when the values are ranked. Thus the median rainfall at Dakar is 475 mm, that is, rather less than the average of 550 mm. In many ways, the median provides a more accurate picture of 'average' precipitation or a more reliable expression of variability. This is because the median value is more likely to occur in any one year than the so-called average.

(c)   Maximum and minimum values

A simple way of expressing monthly or yearly rainfall variability is to quote the maximum and minimum values in relation to the mean. At Shannon the annual maximum rainfall is 1178 mm, and the annual minimum is 706 mm. These figures represent a range of 76–127% of the mean. At Dakar, however, the annual minimum value is 23%, whereas the annual maximum is 167% of the mean. These data suggest, once again, that annual rainfall is much more variable, and therefore uncertain, at Dakar than at Shannon.

(d)   Relative variability

The deviations of the individual values of rainfall of a frequency distribution from the mean can also be summarised by calculating the *mean deviation*. This is the sum of all deviations, without respect to sign, divided by the number of observations.

Thus the mean deviation at Dakar is 152 mm, whereas it is only 101 mm at Shannon. When the mean deviation is expressed as a percentage of the average, the *relative variability* of rainfall is obtained. The mean deviation at Shannon is thus 11% of the mean value, indicating low

Figure 6.11   World map of annual relative rainfall variability. (Source: Trewartha and Horn, 1980, p. 213)

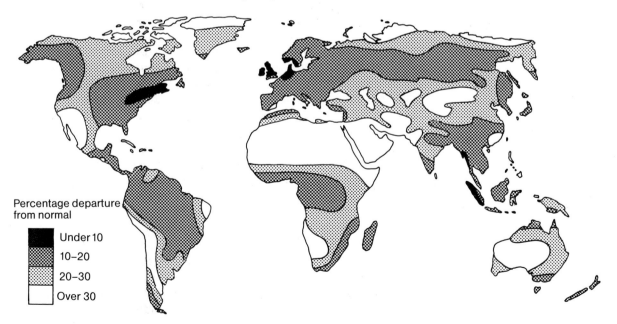

Percentage departure from normal

Under 10
10–20
20–30
Over 30

relative variability. By way of contrast, the mean deviation is 28% of the mean at Dakar. Thus, the substantially higher relative variability of rain at Dakar once again becomes apparent.

A number of studies have suggested that, with few exceptions, the relative variability of mean annual rainfall is inversely related to the average annual amounts: that is, stations with the greatest total volume of rainfall have the smallest annual percentage variation. This relationship is largely confirmed by comparing Figures 6.2 and 6.11. Relative variabilities are very low in the British Isles and in Central Africa, both areas having fairly high rainfall totals. More generally, a variability in the annual amount of rainfall of less than 15% is typical in the high rainfall areas of the Atlantic and Pacific depression belts and in much of the equatorial region. In the semi-arid areas of the world, with annual rainfall totals between 300 mm and 500 mm, relative variabilities are in the region of 20–25%. This means that the annual reliability of rainfall is much reduced in these areas and prolonged dry spells may occur. Relative variability values in excess of 40% are typical for the deserts of Africa, Arabia and Asia, and along the west coast of South America. In the interior of these deserts, annual rainfall variability or irregularity is very high.

2. Time-series variability

A major limitation of the annual or monthly variability statistics discussed above is that they do not tell us anything about rainfall trends or, in other words, the time-sequence of rainfall occurrence. The importance of time-series analyses in rainfall supply will now be demonstrated with respect to long-term trends and short-term fluctuations.

(a) Long-term trends

A feature of variability in most regions is for groups of dry or groups of wet years to occur together. This situation is well illustrated at Ed Dueim, a station with a semi-arid climate in eastern central Sudan. It is possible to identify three separate rainfall trends for this station (see Figure 6.12). First, there is a very varied annual rainfall sequence of above and below average rainfall. Second, significant dry spells are more frequent throughout the 1940s and in the late 1960s and 1970s, whereas wet spells are more characteristic of the 1930s and 1950s. Third, as illustrated particularly by the five-year running mean, there is an overall decrease in annual rainfall amount between 1935 and the mid-1970s. The prolonged dry spell of the late 1960s to the mid-1970s has had particularly serious consequences for Ed Dueim, as indeed for most of the Sahel region of sub-Saharan Africa. Between 1968 and 1974, annual precipitation in the Sahel (Mali, Niger and Upper Volta) was consistently not more than one-half of the long-term average. In Senegal, very dry conditions occurring in 1973/4 can be traced back to low rainfall in 1968, 1970, 1971 and 1972 (see Table 6.1). The better-than-average rainfall year in 1969 was not sufficient to reverse or contract the intensification of aridity and drought in the region.

Figure 6.12   Annual rainfall trends at Ed Dueim in central Sudan. (Source: Shakesby and Trilsbach, 1982)

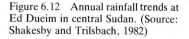

Table 6.1 Mean annual rainfall and annual rainfall at Podore and Linguere in Senegal, 1966–72

Station	Mean annual rainfall (mm)	Annual rainfall (mm)						
		1966	1967	1968	1969	1970	1971	1972
Podore	335 (1940–69)	243	271	184	432	254	136	80
Linguere	506 (1937–69)	504	556	279	679	295	328	219

Source: Jackson, 1977

(b) Short-term fluctuations

Expressions of annual and even monthly variability and reliability suffer the same kind of limitation as annual averages, in that they give no indication of shorter period conditions, and the precise timing of rainfall events. This is an important omission, because, if anything, rainfall variability increases over shorter time-scales.

In the southern Sudan at Abu Deleiq, a station with a tropical or seasonal wet and dry regime, rainfall records have been kept since 1905. The shortest rainy season here lasted from 28 August to 15 September (in 1917) and the longest from 18 March to 28 October (in 1928). As shown in Figure 6.13 for Yundum Airport in the Gambia, a high daily or weekly variation in rainfall suggests that defining the 'start' (and 'finish') of the rains may not be straightforward. Figure 6.13 shows that the onset of the rainy season is not easy to demarcate, and varies widely from year to year. Moreover, dry spells of various lengths are surprisingly frequent, even during the rainy season. In Singapore, daily rainfall totals for 1963 are shown in Figure 6.14 and emphasise how irregular the pattern can be.

Daily rainfall variability often renders average monthly and annual values impractical. For instance, at Doorbaji in the Thar Desert, where the

Figure 6.13 Occurrences of daily rainfall (≥ 12.7 mm) at Yundum Airport, the Gambia, West Africa, 1947–82. Recommended notional planting date for ground-nuts, the major crop, is 7 July. This crop requires 120 days from planting to maturity, which is the length of the growing (wet) season in the average year. (Source: Hutchinson and Sam, 1984, p. 25)

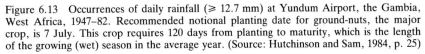

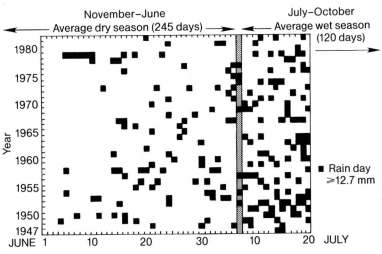

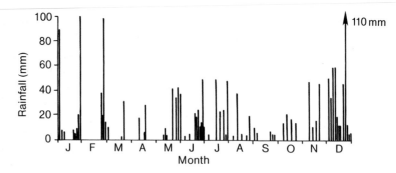

Figure 6.14 Daily rainfall at Singapore in 1963. (Source: Jackson, 1977, p. 63)

average annual rainfall is 125 mm, a rare storm gave 850 mm over two days. At the Red Sea station of Hurghada (Egypt), with a mean annual rainfall of about 3 mm, 41 mm fell on 8 November 1939. Even in wet locations such as Singapore, much of the rain is concentrated in short periods of a few days of heavy rain. The percentage received in the three wettest days of the month reached 78% in August 1963, 90% in March 1963 and 100% in April 1963. Analyses of rainfall variability over fairly short time-scales thus lead to questions concerning rainfall intensity. This topic is pursued below.

3. Rainfall intensity: tropical and temperate environments

Although the annual and seasonal distribution of rainfall exerts an overall control on water availability, and variability imposes a degree of uncertainty, it is often the characteristics of individual rainstorms which are of greatest importance, especially their intensity, duration and frequency. As these three rainfall dimensions are closely related, they are best considered together.

(a) 24-Hour intensity

If the total annual rainfall at any particular place is divided by the number of rain days, the mean 24-hour intensity can be given. Table 6.2 shows that

Table 6.2 Annual mean rainfall per rain day

Station	Rainfall (mm)
Quito, Ecuador	8.5
Georgetown, Guyana	13.3
Accra, Ghana	13.6
Lagos, Nigeria	14.4
Bombay, India	22.4
Rangoon, Burma	20.9
Jakarta, Indonesia	13.5
London, England	5.5
Vienna, Austria	4.1

Rain days are defined here as days with at least 1.00 mm of rain.
Source: Nieuwolt, 1977, p. 122. Copyright John Wiley & Sons Ltd, 1977. Reprinted by permission.

when using the mean 24-hour rain-day intensity, rainfall in tropical areas is more intense than in temperate environments. One possible explanation of this is the higher rainfall totals common in tropical areas.

In tropical and temperate regions alike, however, a large proportion of the total rainfall at any place is accounted for by a relatively small number of rain days. This means that on these rain days rainfall intensities are much higher than the average 24-hour rain-day value. Riehl (1978) claims that in most areas 10–15% of the rain days account for as much as 50% of the rain, whereas, in contrast, 50% of the rain days with the smallest rainfall amounts produce only 15% of the total.

(b) Short-term rainstorm intensity

In temperate areas, the small number of rain days that contribute significantly to total rainfall is characterised by relatively prolonged rainstorms (up to four hours) of moderate to fairly high intensity. In the tropics, on the other hand, most rainfall comes from very intensive convective storms of short duration (up to one hour). Accordingly, when intensity is measured over the period of actual rainfall, the high intensity of tropical rain becomes even more apparent. Figure 6.15 shows that maximum one-hour falls largely occur in the tropics. The highest values of all (over 60 mm/h) are characteristic of the equatorial trough, where convective storms charged with a plentiful supply of moisture are common.

These data support the claim that 45% of the rainfall in the tropics, but only 5% in temperate areas, occurs at intensities of at least 25 mm/h. Above this rainfall rate, rainfall is thought to become erosive. Whereas

Figure 6.15   One-hour precipitation (mm) likely to be exceeded once every two years. (Source: Jackson, 1977, p. 79)

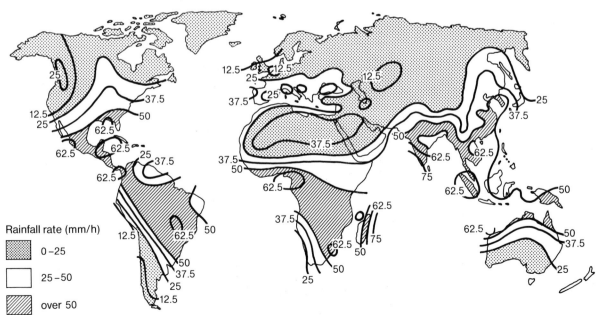

133

in temperate areas intensities rarely exceed 75 mm/h, and then only in very short-lived summer thunderstorms, in the tropics intensities of 150 mm/h occur regularly, and a rate of 340 mm/h for a few minutes has been registered.

ASSIGNMENTS

1. *Examine Table 6.3.*

    (*a*) *Using Figure 6.10 (a) and (b) as a model, construct frequency distribution diagrams of annual rainfall occurrence for Valentia and Nagpur.*

Table 6.3   Annual rainfall at Nagpur, Central India, 1931–60 and annual and June rainfall at Valentia, south-west Ireland, 1940–81

Nagpur, Central India (tropical wet and dry regime)		Valentia, SW Ireland (temperate oceanic west-coast regime)		
Year	Annual rainfall (mm)	Year	Annual rainfall (mm)	Monthly rainfall (June) (mm)
1931	1543	1940	1327	54
1932	1190	1941	1283	21
1933	1905	1942	1070	13
1934	921	1943	1408	137
1935	1102	1944	1261	50
1936	1632	1945	1448	116
1937	1681	1946	1544	108
1938	1502	1947	1551	94
1939	923	1948	1483	94
1940	1605	1949	1265	73
1941	1094	1950	1633	65
1942	1444	1951	1596	52
1943	910	1952	1374	80
1944	1303	1953	1278	58
1945	1316	1954	1548	72
1946	1345	1955	1148	95
1947	1202	1956	1240	66
1948	1156	1957	1447	47
1949	1411	1958	1589	127
1950	731	1959	1433	67
1951	1002	1960	1690	68
1952	703	1961	1375	59
1953	850	1962	1232	69
1954	1204	1963	1395	54
1955	1388	1964	1487	112
1956	1278	1965	1357	134
1957	1049	1966	1429	91
1958	1076	1967	1347	25
1959	1348	1968	1457	78
1960	1249	1969	1131	87
		1970	1393	107
		1971	935	76
		1972	1509	86
		1973	1229	68
		1974	1591	38
		1975	1206	18
		1976	1446	111
		1977	1541	60
		1978	1508	90
		1979	1574	77
		1980	1775	85
		1981	1408	89

(b) *Denote on the diagrams the rainfall average, the median and the maximum and minimum recordings in each case.*

(c) *Compare the variability patterns of the two stations using your results.*

(d) *Comment on the rainfall variability of the two stations in relation to their respective rainfall regime.*

2. (a) *Follow stages 1(a) and 1(b) again, using the monthly rainfall data for Valentia.*

(b) *Examine both the annual and monthly rainfall variability at Valentia.*

(c) *Using your results so far, and other information from the text, examine the evidence which claims that rainfall variability increases as time-scales decrease.*

3. (a) *From Table 6.3 draw a diagram of the annual rainfall trend at Valentia and Nagpur by plotting rainfall totals against time.*

(b) *What aspects of variability are indicated by your diagram which are not included in assignment 1?*

(c) *Explain why the time sequence of rainfall events is an important aspect of rainfall variability.*

C. Environmental and Human Response

As with heat and solar energy, water is a vital requirement for all living systems. From the point of view of life on the planet, precipitation is arguably more important than temperature. As we have seen, precipitation input to the earth/ocean surface is very uneven in space and time. Variations in precipitation across the face of the planet are more considerable than those of temperature, with correspondingly more intense and observable effects on global ecosystems. A failure of the rains, for instance, can have devastating effects on individual ecosystems. This point is graphically illustrated by the droughts of the 1970s and 1980s in the Sahel region of sub-Saharan Africa, and in 1976 and 1984 in the United Kingdom.

Human and environmental systems adjust to variations in precipitation in different ways. First, there are adaptations to regular seasonal patterns of variability, including the demands of alternating wet and dry seasons. Second, difficult adjustments have to be made in coping with irregular drought. Finally, variations in rainfall intensity impose on ecosystems a number of adjustments in relation to problems of soil erosion and flooding.

1. Adaptations to regular rainfall events in West Africa

In West Africa, the length and severity of the dry season is one of the most important factors influencing agricultural patterns with respect to both arable and livestock systems.

In the humid tropics, where there is both abundant rainfall (over 2000 mm) and little or no dry season (for example, Monrovia, Axim), cultivation can take place the whole year round (see Figures 6.16 and 6.17). Those crops flourishing under continuous hot wet conditions and

Figure 6.16 Idealised model of the relation between climatic type (rainfall regime), vegetation formation and cropping pattern between the equator and the tropic of Cancer. West Africa is taken as an example. (Source: Manshard, 1979, p. 23)

Station	Axim Monrovia	Enugu	Minna	Sokoto	Ménaka	Kidal
Number of humid (or dry) months	10–12 (0–2)	9–10 (2–3)	7–9 (3–5)	4–6 (6–8)	1–3 (9–11)	0 (12)
Mean annual rainfall (mm)	Mainly > 2000	Mainly >1500	Mainly > 1000	750–1000	>400 < 400	< 250
Typical economically useful plants	Rubber, tropical timbers	Oil palm, cocoa, coffee	Yams, maize	Cotton, millet, ground-nuts	Ground- nuts	
Simplified transect sketch 						
Main vegetation	Moist tropical forest	Monsoon forest	Wet savanna	Dry savanna	Thorn-bush savanna Semi-desert	Desert

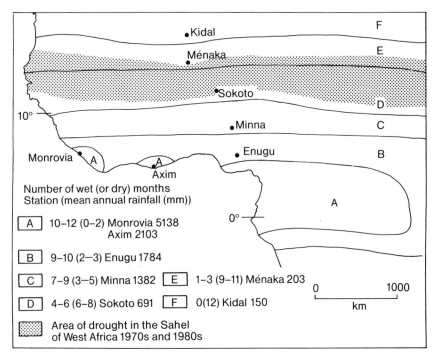

Figure 6.17 Map of the average duration of the wet (and dry) season in West Africa in relation to the area affected by drought in the Sahel in the 1970s and 1980s

Number of wet (or dry) months
Station (mean annual rainfall (mm))

A 10–12 (0–2) Monrovia 5138 Axim 2103

B 9–10 (2–3) Enugu 1784

C 7–9 (3–5) Minna 1382 E 1–3 (9–11) Ménaka 203

D 4–6 (6–8) Sokoto 691 F 0(12) Kidal 150

Area of drought in the Sahel of West Africa 1970s and 1980s

not needing a pronounced dry season for harvesting are grown. Root crops such as yams, sweet potato and cassava are often the most important subsistence crops (that is, crops grown for home consumption). Tree crops of commercial significance include rubber and oil palm, and tropical timbers may be harvested. Excessive rainfall can cause the removal of nutrients from soils by leaching, rendering the maintenance of soil fertility difficult.

In areas where rainfall totals remain high (over 1500 mm), but where there is a short dry season of 2–3 months (for example, Enugu), perennial tree crops (oil palm, coffee, cocoa) are of major importance, together with food crops such as yams and cassava. Two cropping seasons per year are possible, the first being usually one of the aforementioned food crops (yams) or cash crops (coffee), the second a crop such as cotton, requiring a short dry period. Where the dry season lengthens to 3–5 months (for example, Minna), perennial tree crops decline in significance, but yams, cassava and grain crops such as maize, which are able to ripen under the short sunny dry period, do well.

As the dry season increases further in length and severity, it may be possible to grow only one crop per year, unless there is access to irrigation. A pronounced dry season of 6–8 months with moderate rainfall amounts of 750–1000 mm (for example, Sokoto) necessitates the cultivation of drought-resistant crops including millet, sorghum, cotton and ground-nuts. Perennial crops resistant to drought such as sisal, cashew nuts and tung oil may be grown. In very dry areas with annual rainfall totals of less than 500 mm and with a long dry season of 9–10 months (for example, Ménaka), perennial crops are not suited, but short-term drought-resisting annuals may just be grown, including millet, ground-nuts and sesame.

Cattle-rearing in the hot, more humid parts of West Africa is greatly restricted by the presence of the tsetse fly. This insect transmits the disease of trypanosomiasis (sleeping sickness) to both humans and livestock. It is only in those areas with a significant dry season of six months or more (that is, in the dry savannas and semi-arid zones) that any significant numbers of cattle can be kept, usually by nomadic pastoralists. Many nomadic pastoralists migrate with their animals, however, into the wet savannas when the dry season prevails there, and move back to the dry savanna as the wet season approaches.

2. Adjustment to irregular rainfall events

(a) The concept of drought

Drought is a term difficult to define precisely, but implies an absence of precipitation for a period which is long enough for moisture deficits to develop in the soil. Such moisture deficits are the result of water loss by evapotranspiration, ground water and stream flow. Although drought clearly involves a reduction of precipitation and water shortage, it is not synonymous with soil moisture deficiency, although the two concepts are related. For instance, during the dry season in a tropical wet and dry climate, there may be severe soil moisture deficiency, but a condition of drought may not be perceived by the inhabitants. This is because the dry season is a regular occurrence and does not disrupt normal biological and human activity. In other words, drought is related to an *unexpected* failure of the rains at a particular time, since most activities using water will be geared to that which is normally available.

(b) Drought adjustment

The severe droughts which affected the UK in 1976 and 1984 remind us that drought is by no means restricted to arid and semi-arid areas. Nevertheless, drought and crop failure are a common feature in areas of low and variable rainfall. Today, no less than 630 million people, or 14% of the world's inhabitants, live in arid and semi-arid environments. Of these, some 50 million people are constantly faced with malnutrition and possibly death whenever the rains fail.

(i) *Short-term drought response.* In Europe and north-east USA, where agriculture and other activities are geared to rainfall received at fairly regular intervals, and where user demand is high, a failure of the rains for two or three weeks may induce a condition of drought. In the UK, for example, the Meteorological Office defines an absolute drought as a period of at least 15 consecutive days with less than 0.2 mm of rain per day. As already implied, such definitions are of little practical use in seasonal rainfall areas in the intertropical zones, which normally experience many successive dry months. Nevertheless, short-term droughts can occur in these areas, when the rains fail at a critical period during the rainy season (see Figure 6.13). A delay of even a week or two in the onset of the rains can destroy all hopes of a normal harvest. Such droughts may, first, foreshorten the growing season and compress the agricultural workload. Second, if fodder is scarce at the end of a long dry season, short delays in the arrival of the rains can mean serious animal losses. In contrast, in the middle of the rainy season, not only might rainfall occurrence be expected to be more certain, but soil moisture reserves will be able to cope with any absence of rain for short periods. However, towards the end of the wet season, when a greater chance of rain failure can be expected again, crops may be forming their yields of fruit or grain, and thus be in need of an adequate supply of moisture.

(ii) *Long-term drought response.* Rainfall deficiency over a large part or all of the growing season, or over several years, has more dramatic repercussions for individual ecosystems. In the arid and semi-arid climates, where rainfall is barely adequate for normal agricultural use and where rainfall variability is high, such season and year-long droughts are not uncommon. In poor agricultural societies in these areas, a departure from the annual mean of 25% will injure crops, and a departure of 40% will cause widespread crop failure and famine.

It may be recalled from section B of this chapter that an important aspect of rainfall variability is the tendency for sequences of dry (or wet) years to occur. One exceptionally dry year (or wet one, creating problems of flooding and soil erosion) can be survived, despite its tragic effects. It is more difficult, however, for the peasant cultivator with few resources to survive than, say, a plantation system or the highly capitalised systems of temperate areas. The real problems arise when such conditions recur over a series of years; the peasant cultivator is, again, especially hard hit.

(c) Case-study: the Sahel drought

(i) *Climatic versus anthropogenic factors.* As previously implied (page 130), recent drought in the Sahel is more correctly seen as the culmination of a number of poor rainfall years rather than as a single disastrous drought year. During the Sahel drought of 1973, a failure of the rains since about 1968 resulted in almost total crop failure in that year. Since most of the Sahelian inhabitants are nomadic pastoralists, great numbers of livestock succumbed to the drought. It has been estimated that livestock losses must have totalled several million, including an 80% loss of cattle. More serious still, about 200 000 people died in the Sahel in 1973 alone, and about 7 million people became dependent on food aid by national and international agencies. This aid and the drought continue today, especially in Somalia and Ethiopia.

It is important to bear in mind that the effects of such a long-term drought can be amplified by economic, social and political factors (see Barke and O'Hare, 1984). Many people believe that human misuse of land in arid and semi-arid areas is a major factor in the process of desertification. *Desertification* may be defined as the creation of desert-like surfaces by a reduction in the quality of soils and vegetation (Plate 6.1). The desertification risk is greatest when a succession of good years with above average rainfall is followed by a long drought (see Figure 6.12). The years of favourable rain encourage the inhabitants of the area to expand the cultivated areas, to build up livestock numbers and to increase

Plate 6.1 Desertification (Photograph: Audio-Visual Productions)

their own population. When the rains fail during the ensuing drought, the carrying capacity of the land to support such activity is quickly exceeded. Soils and vegetation, already under severe stress from lack of rainfall, are impoverished further by overgrazing and compaction by livestock, woodland clearance for firewood, and over-cultivation. In this way, hundreds of thousands of hectares of grazing land along the southern and northern fringes of the Sahara are being converted to desert each year. Thus, instead of the desert being pushed outward by climatic influences, it may be more correct to suggest that it is being pulled out by human mismanagement of the semi-arid zone.

(ii) *Adaptation and prevention.* There are three different ways of facing up to the problem of long-term drought (Whittow, 1980). First, one can bear the loss and do nothing. Second, it may be possible to move away from the drought-stricken area and settle elsewhere. In West Africa, cultivators and herdsmen trekked southwards into the more humid savanna zones in large numbers. This resulted, however, in increased land pressure in the receiving areas. Third, one can reduce the impact of the drought hazard by adopting remedial measures.

Of the remedial measures, there are three possible approaches. First, the scientific approach deals with the modification of the hazard itself. This involves conscious efforts to increase rainfall in the drought areas by cloud-seeding. Another attempt is the 'green-belt' solution, whereby massive tree and shrub-planting programmes are designed to improve the overall water balance of the region, by increasing atmospheric humidity (through enhanced evapotranspiration). On a large scale, there are doubts about the expense and effectiveness of both these methods.

A second method is the technical approach. This considers the possibility of irrigating the Sahel on a large scale by using and developing new sources of water. Large new irrigation schemes based on river water, or ground water from artesian basins, are likely to prove expensive on a large scale.

The third possibility, and the one which has received major emphasis in recent years, is the behavioural approach. This method includes attempts to change peoples' customs, habits and values. Strategies here involve a number of closely related issues, such as reducing the size of cattle herds; the need for strict grazing control; the re-settlement of some herdsmen in permanent dwellings; small-scale irrigation projects; the use of new sources of energy (for example, solar) and energy conservation in cooking to offset the effect of firewood gathering; and land reform to redistribute land held by a few wealthy owners to the large number of peasant cultivators and herdsmen. Last, but not least, a plea is made for population reduction through family planning programmes.

As with the scientific and technical approaches, many of these suggestions will not be easy to implement, although they seem superficially attractive and low cost. For instance, the desert nomads own large numbers of animals (twice as many as the Sahel can normally be expected to carry) not for economic profit, but for social and political prestige, and as an insurance against drought. If the rains fail, some animals can be sold, some will die, but there may not be a total loss.

3. Rainfall intensity: environmental adjustment

Rainfall intensity has a considerable bearing on the effectiveness of rainfall for agriculture, and on the rate of soil erosion and flooding.

(a) Rainfall effectiveness

Effective rainfall is that fraction of the total entering the soil and remaining within the root zone. That lost as surface evaporation and runoff, and that draining beyond the root zone is ineffective. Rainfall intensity has an important influence on rainfall effectiveness, although other factors – including slope, the presence or absence of vegetation, soil infiltration capacity and storage, and evaporation conditions – make a contribution.

Rainfall is less effective in the tropics than in temperate latitudes, because of the higher intensities in the former area. The high intensity of tropical rain results in a good proportion of surface runoff and little in the way of soil infiltration. Second, in tropical regions of marginal rainfall, there may be considerable intervals between rainstorms, when the soil may dry out, bake and harden, and thus offer little scope for surface absorption. Third, in tropical semi-arid areas, a general absence of vegetation cover restricts infiltration of water via the root zone and presents no obstacle to check rapid runoff of water on slopes.

(b) Soil erosion

A much higher proportion of tropical than temperate rainfall occurs at intensities which are thought to be erosive, that is, at 25 mm/h and above. Taken with the fact that rainfall totals are often higher in the former area than in the latter, the far greater energy available for erosion in tropical latitudes becomes apparent.

In the tropics, much of the rainfall results from a comparatively small number of intense storms. It follows that much of the erosion should occur in a limited period of time. At Mazoe, in Botswana, more than half the erosion occurred during the one or two heaviest storms of the year. In one case three-quarters of the yearly loss took place in 10 minutes. It has also been shown that almost 90% of the annual soil erosion in Upper Volta occurred in one 14-hour storm in 1956 and one 6-hour storm in 1957.

ASSIGNMENTS
1. *Consult Figures 6.16 and 6.17.*
    (a) *Describe the regional variation in the nature of the dry season in West Africa.*
    (b) *Using the text and relevant diagrams, examine the factors that produce such variation.*

(c) *Describe and explain the relationship between agricultural and climatic patterns in West Africa.*

2. (a) *Define the terms* drought *and* desertification.

   (b) *Examine the climatic and anthropogenic factors that have been responsible for desertification in the Sahel.*

   (c) *Give a reasoned account of the methods which may be adopted to alleviate the impact of desertification.*

3. *Compare the effects of rainfall intensity in temperate and tropical environments.*

Key Ideas

Introduction

1. A chief feature of rainfall is its variation in space and time.
2. Traditionally, geographers have tended to study long-term aspects of variability, but studies are increasingly being made of short-term events.

A.   World Annual Rainfall Distribution

1. There is a dominant zonal or north–south arrangement in the global mean annual distribution of precipitation.
2. High precipitation totals are found at the equator and in the mid-latitudes; low totals in the subtropics and in polar regions.
3. The annual zonal pattern is comprised of strong seasonal swings in the distribution of rainfall.
4. Zonal and seasonal patterns at the global level can be partly explained using the model of the general circulation.
5. Significant longitudinal or east–west variations in global mean annual precipitation interrupt the simple zonal model.
6. Temperature and pressure contrasts between land and sea, together with ocean currents and major relief barriers, are responsible for producing regional variations in rainfall.
7. An understanding of precipitation-creating mechanisms at the global level allows eight major precipitation zones to be identified, together with a number of other regimes produced by the modification of the zonal arrangement.

B.   Rainfall Variability

1. Rainfall variability is an expression of the regularity of rainfall occurrence in time.
2. Rainfall variability can be denoted with the use of frequency distribution diagrams.
3. These diagrams allow rainfall distribution to be expressed in terms of the annual range, variation from the mean and median values, and relative variability.

4. In general terms, there is an inverse relationship between annual rainfall amount and relative annual variability.
5. The time sequence of rainfall occurrence, such as long-term trends and short-term fluctuations, is an important aspect of variability.
6. Rainfall intensity, on a daily or shorter-term basis, is much higher in tropical than in temperate areas.

C. Environmental and Human Response

1. Human and environmental systems are affected by regular seasonal changes in rainfall distribution; unexpected or irregular changes in rainfall occurrence; and variations in rainfall intensity.
2. There is a close relationship between agricultural and seasonal precipitation patterns in West Africa.
3. Drought is associated with soil moisture deficiency, caused by an unexpected failure of the rains over long or short-term scales.
4. Desertification is the creation of desert-like surfaces in the subhumid and semi-arid areas of the world from a reduction in the quality of soils and vegetation.
5. Both climatic drought and poor land management can accelerate the process of desertification.
6. Scientific, technical and behavioural strategies can reduce the impact of desertification, but their implementation faces many difficulties.
7. Rainfall intensity, especially in the tropics, has a marked influence on rainfall effectiveness and on the rate of soil erosion and flooding.

Additional Activities

1. (a) Using tracing paper, and Figure 6.2, map the world distribution of the major deserts (that is, areas whose rainfall is less than 250 mm per annum).
    (b) Refer to Figures 2.8 and 2.9. Summarise the major deserts in relation to their seasonal temperature conditions, that is (i) hot deserts (warm in winter, above 15 °C, hot in summer, above 30 °C); (ii) cool deserts (cold in winter, below 0 °C, hot in summer, about 20–30 °C); and (iii) cold deserts (very cold in winter, well below 0°C, cool in summer, about 10–15°C).
    (c) Using Figure 2.5, categorise the various deserts in relation to solar radiation receipt.
    (d) Account for the global distribution of deserts using one or more of the following: (i) the influence of the subtropical highs, (ii) rain shadow and adiabatic warming effects; (iii) cold currents; (iv) distance from rain-bearing systems and the sea; (v) polar high-pressure systems.
    (e) In what other ways can deserts be created?
2. (a) Construct graphs of the data shown in Table 6.4. We recommend that you use a bar graph for monthly rainfall between January 1975 and December 1976, and a line graph to indicate average conditions.

Table 6.4   Rainfall (in mm) for 1975 and 1976 and average monthly rainfall, 1941–70, for England and Wales, and Scotland

	J	F	M	A	M	J	J	A	S	O	N	D	Year
A. England and Wales													
1975	117	31	81	71	47	21	66	52	107	36	73	50	752
1976	60	40	43	21	64	17	32	27	160	153	83	94	794
Average (1941–70)	86	65	111	86	67	62	73	90	83	84	97	89	906
B. Scotland													
1975	245	48	58	100	48	67	112	86	184	78	128	90	1244
1976	185	87	130	60	118	65	64	25	141	202	127	110	1314
Average (1941–70)	137	104	92	90	91	92	112	128	137	150	142	155	1430

Source: Doornkamp and Gregory, 1980

Figure 6.18   Rainfall distribution in the UK, April–August 1976 as a percentage of the April–August 1916–50 average. (Source: Roy, Hough and Starr, 1978, Figure 2)

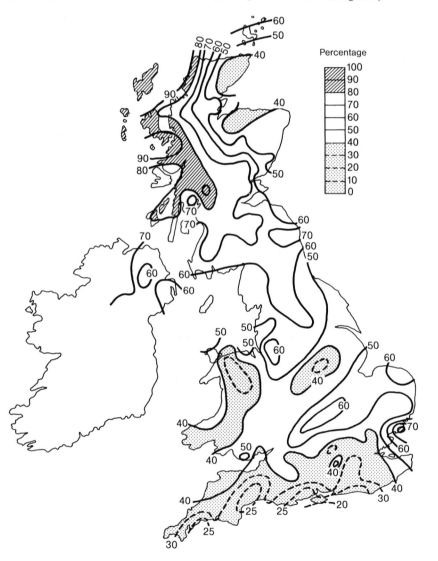

Figure 6.19   Weather map of 26 August 1976, showing meteorological conditions at the height of the British drought

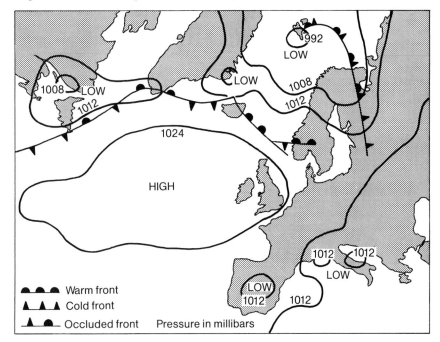

Figure 6.20 Upper air conditions during the 1976 British drought, showing the influence of a blocking high on the direction of upper air movement (i.e. the jet stream) and thus on the passage of surface weather system (e.g. depressions). (Source: Doornkamp and Gregory, 1980)

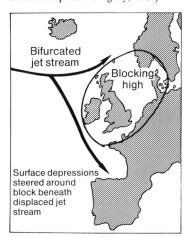

(b)   Describe the patterns of drought shown.

(c)   Describe the moisture patterns shown in Figure 6.18.

(d)   Examine Figures 6.19 and 6.20. Explain the links between the synoptic situations shown and the meteorological conditions indicated in Table 6.4 and Figure 6.18.

3.   Discuss the relationships between precipitation patterns and human activity in tropical environments.

7 Weather Systems at the Synoptic Scale

A. Air Masses and Fronts

1. Air masses

It will be recalled from Chapter 2 that the atmosphere is warmed primarily by the interception of outgoing long-wave radiation from the surface. In certain circumstances, therefore, it is reasonable to expect the temperature and moisture-content characteristics of air over a region of the earth's surface to reflect conditions below. For example, the air overlying a warm, moist surface might be expected to be warm and humid by comparison to that over a cold, dry surface. Marked geographical diversity at the surface would obviously induce considerable variation in the air aloft, whereas a homogeneous surface would tend to promote the development of bodies of air aloft with fairly uniform temperature and humidity at similar heights. Such bodies of air often exhibit only very gradual changes in these properties over considerable horizontal distances, perhaps thousands of kilometres, and are termed *air masses*. The areas over which they form are known as *source regions*.

(a) Source regions

Source regions are usually places where only minimal changes in surface geography occur over large expanses, for example, ocean surfaces, ice-covered areas, deserts, large plains. These will be particularly effective areas for forming air masses if the air above them is slow moving or gently subsiding. This gives time for the transmission of surface-related characteristics to proceed. Areas of stationary or slow-moving anticyclones are ideal in this respect, and almost all air masses develop in regions where these predominate.

(b) Classification

Air masses are labelled according to two criteria. First, with respect to moisture content, those that have originated over an ocean surface are designated m (maritime), whereas those formed over a land surface are labelled c (continental). Second, an indicator of their thermal characteristics is derived on a latitudinal basis: A or AA (Arctic or Antarctic), P

Table 7.1 Properties of major air masses

| | Characteristics at source | | |
	Temperature (°C)	Specific humidity (g/kg)	Typical properties
Maritime tropical (mT)			
summer	22–30	15–20	mild, moist
Maritime polar (mP)			
winter	0–10	3–8	cool, moist
summer	2–14	5–10	cool, moist
Continental tropical (cT)	30–42	5–10	warm, dry
Continental polar (cP)			
winter	−35–−20	0.2–0.6	cold, dry
summer	5–15	4–9	cold, dry
Continental Arctic (cA)			
winter	−55–−35	0.05–0.2	very cold, very dry
Maritime equatorial (mE)	approx. 27	approx. 19	warm, very moist

(Polar), T (Tropical) or E (Equatorial). Table 7.1 summarises the properties of six important air masses, which can be identified using these criteria in combination.

(c) Influence and interaction

When an air mass moves out from its source region, although it tends to undergo progressive modification, it brings its distinctive properties with it to influence weather at distant locations. The climate of a place may thus be described in terms of the frequency with which it experiences particular air masses. Figure 7.1 shows how seasonal changes in temperature and precipitation may be explicable in terms of the changing influence of certain air masses. A place such as Farina, under the influence of the same air mass throughout the year, shows little seasonal change, whereas Darwin has a regime involving an oscillation between two different air masses.

The planetary wind circulation system makes interaction between air masses more likely to occur in some areas. For example, in Figure 7.2 a

Figure 7.1 Air mass domains and monthly temperature and precipitation averages for Darwin and Farina, Australia. (Source: Oliver, 1973, Copyright John Wiley & Sons, Ltd)

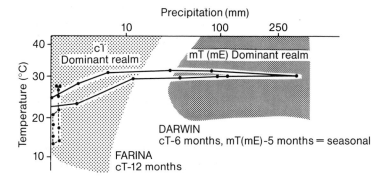

Figure 7.2 Air mass source regions. (Source: Oliver, 1973, Copyright John Wiley & Sons, Ltd)

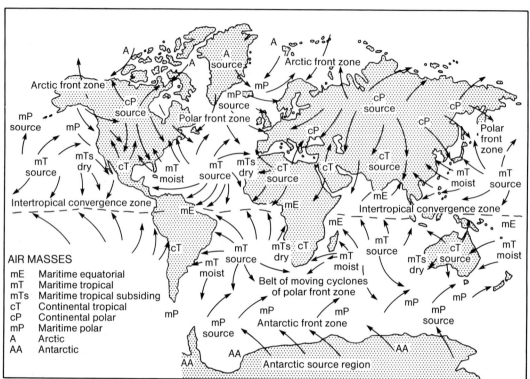

convergence of air in the vicinity of the equator is indicated. Since the source regions from which these air streams originate, however, are at similar latitudes in either hemisphere, both air masses will have similar properties and will mix freely. This is in sharp contrast to the other main area where air-mass interaction is indicated – the zone where tropical and polar air masses collide at the high–middle latitudes in each hemisphere. In this case, fundamentally different air masses are coming into conflict along a boundary zone which has become known as the *polar front*.

2. Fronts

When two fluids with different densities come together they do not readily mix (for example, oil and water). Similarly, two air masses with different temperatures, and consequently different densities, also resist mixing and a transitional zone between them – known as a *front* – becomes established. The term dates from the First World War, when a group of Norwegian meteorologists saw the contest for mastery between two contrasting air masses as akin to the trench warfare on either side of the 'western front'.

Atmospheric fronts are typically 100–200 km wide and are zones of steep horizontal temperature gradients (Figure 7.3). When the cold air is advancing and, because of its greater density, undercutting the warm

Figure 7.3 Temperature characteristics of a frontal zone. (Source: Trewartha and Horn, 1980)

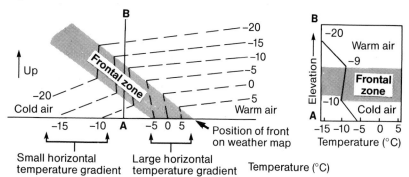

air mass ahead of it, the discontinuity between them is known as a *cold front*. Its passage at the surface heralds the arrival there of the cold air mass. Note that, because of the slope of the front aloft, there is still warm air at height after the passage of the cold front at the surface. Typically, the cold front has a gradient of about 2°. As the advancing cold air forces the warm air in front of it to rise, this produces precipitation in the manner described in Chapter 5. When the warm air mass is advancing, however, because it is less dense, it tends to glide up over any wedge of colder air in its path, producing a *warm front*. Although the gradient of the frontal surface in this case is often below 1°, the cooling of the warm air mass as it ascends is frequently sufficient for cloud formation and precipitation.

Frontal surfaces can thus be identified on satellite photographs by linear cloud formations, which delimit the atmospheric battleground between contrasting air masses (Plate 7.1). This contest is fiercest in the vicinity of the polar front, and produces large-scale disturbances known as *depressions* (in North America these are called *extratropical cyclones*).

ASSIGNMENTS

1. *Table 7.2 shows typical temperature and humidity values for four air masses which affect the British Isles in winter. Giving reasons, identify mP, cP, mT, and cA.*

2. *Give a reasoned explanation of how you would identify fronts (i) on an isotherm map and (ii) on a satellite photograph.*

Table 7.2 Temperature and humidity values for four air masses affecting the British Isles in winter (see assignment A1)

		1000 mb	850 mb	700 mb	500 mb
(i)	Temperature (°C)	1	−9	−20	−40
	Humidity (g/kg)	3.1	1.7	0.7	0.6
(ii)	Temperature (°C)	−2	−12	−22	−41
	Humidity (g/kg)	2.6	1.5	0.6	0.1
(iii)	Temperature (°C)	8	1	−9	−27
	Humidity (g/kg)	5.8	4.0	2.1	0.6
(iv)	Temperature (°C)	11	6	−2	−17
	Humidity (g/kg)	6.8	5.6	3.5	1.2

Plate 7.1 NOAA-7 infra-red 1516 GMT, 1 January 1984. The wide band of stratus running west to east across the centre of the photograph marks the position of the polar front. To the north of it, extensive cumulus development is apparent as cold air flows south over the warm Atlantic Ocean. To the south of the front, cloud development is less marked in the maritime tropical air mass, where gentle subsidence may be occurring. (Photograph: University of Dundee)

B. Disturbances in the Mid-latitude Circulation

In this section depressions and anticyclones are described in detail, before the relationship between upper-air circulation and weather systems at the surface is examined.

1. Depressions

(a) Nature and origin

The convergence of contrasting air masses along the polar front encourages rising air motions, and this leads to a fall in surface air pressure. If conditions in the upper atmosphere are favourable (see p. 158), this flow of ascending air becomes organised into a spiral of upward and inward-moving air known as a *depression*. The anticlockwise flow around this vortex (clockwise in the southern hemisphere) moves polar air further towards the equator on its western flank and drives tropical air further polewards on its eastern flank, producing a bulge or wave on the polar front. Further uplift of air produces more cooling and condensation, resulting in the release of large amounts of latent heat energy. The latter promotes further instability and the disturbance grows rapidly in extent and intensity. Throughout the time this wave depression is developing in amplitude into a mature cyclonic disturbance, it is moving from west to east under the influence of the upper westerlies within which its circulation is embedded.

(b) The life-cycle of a depression: a case-study

(i) *Youth.* Plate 7.2 shows a wave depression forming to the south of Iceland. The northward bulge of tropical air is indicated south-west of the warm front, extending from the peak of the wave towards Ireland. Much of southern Britain and Ireland is covered by thin, low-level cloud, shown by the rather dark grey tones. The brightest tones are seen around the wave disturbance, indicative of thick, high, active cloud formations. These are also well displayed in the vicinity of the cold front part of the wave, trailing south-westwards towards another smaller amplitude wave forming in mid-Atlantic.

(ii) *Maturity.* Considerable development of the wave disturbance took place over the following 32 hours (Plate 7.3). The wave has by now become a vigorous rotating disturbance, centred just north of the Shetland Islands. The cloud band ahead of the warm front lies over Sweden, where precipitation is probably occurring. Less cloud is in evidence to the west of the warm front, where, in the warm sector of maritime tropical air, only a limited amount of uplift and condensation is occurring. Over the southern North Sea, for example, skies are cloudless. The edge of the advancing maritime polar air, in contrast, is well defined by the frontal cloud band running from southern Norway westwards over the Scottish border and Ireland. Behind this cold front, the speckled appearance of the low-level clouds is due to cumulus formation as the cold north-westerly air stream is heated from below as it comes into contact with the warm North Atlantic Drift.

Figure 7.4 shows a diagrammatic representation of a mature mid-latitude depression. Isobars and isotherms indicate the typical pressure and temperature changes that would be experienced as such a depression passed from west to east. Changes in wind direction and strength, cloud

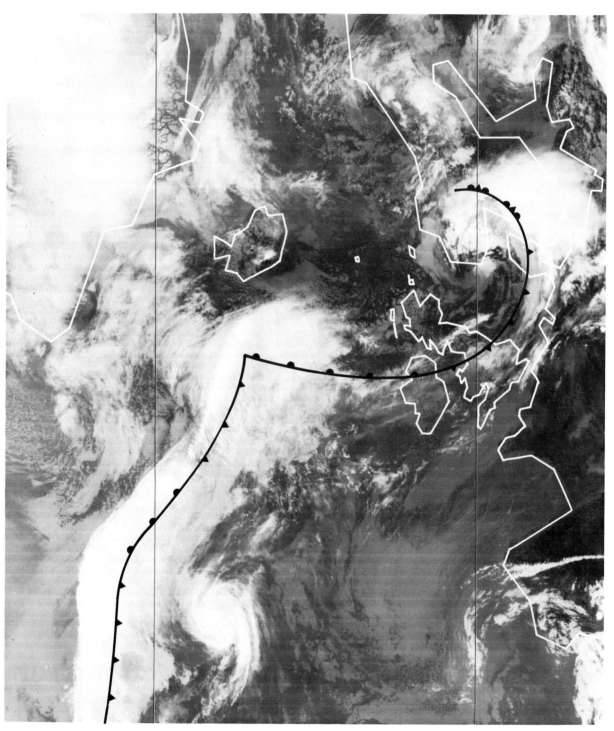

Plate 7.2   NOAA–5 infra-red 1102 GMT, 15 September 1978. (Photograph: University of Dundee)

Plate 7.3   NOAA–5 infra-red 1941 GMT, 16 September 1978. (Photograph: University of Dundee)

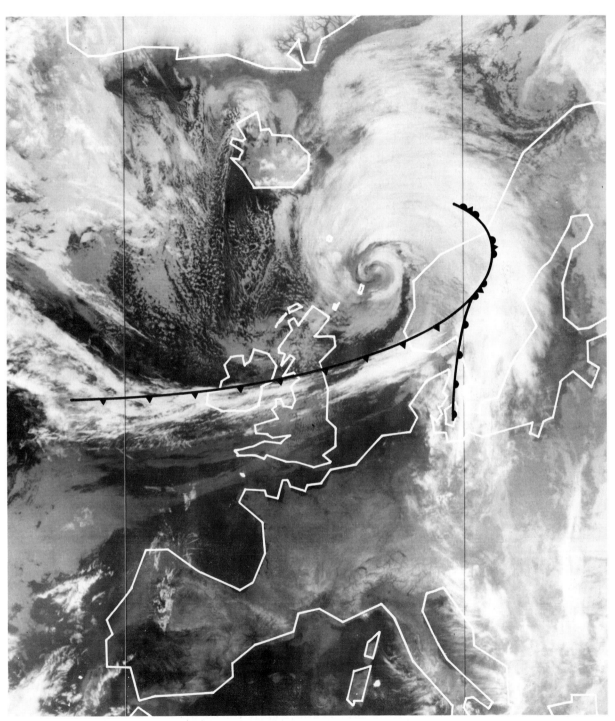

Figure 7.4 Structure of a mature mid-latitude depression. (Source: Trewartha and Horn, 1980)

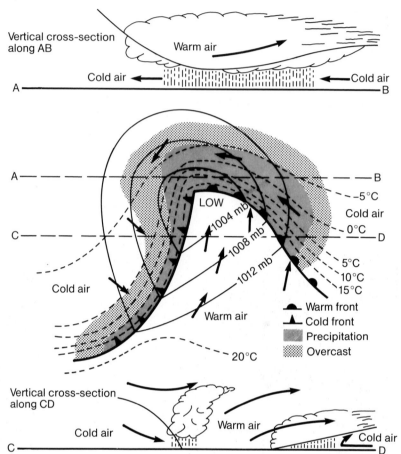

type and height, rainfall and sunshine amounts are also indicated from consideration of a cross-section through the depression, for example along the line CD. Associated changes in humidity and visibility may also be inferred.

(iii) *Decay*. Along the line AB in Figure 7.4, it can be seen that the cold air to the rear of the depression has caught up with the cold air in advance of it, lifting the maritime tropical air clear of the surface. Such a situation is known as an *occlusion*, and is marked on weather charts by a combination of the warm and cold-front symbols. In Plate 7.3 the occluded front, where the two cold air masses come into contact, swings around towards the west from just north of Oslo, into the vortex of the depression. The process of occlusion is a normal stage in a depression's life-cycle, and within 12–24 hours of it commencing, the depression begins to decline in intensity. Since occlusion results in a less vigorous uplift of the warm air mass (the two cold air masses on either side of an occluded front are often at similar temperatures), condensation and hence the release of latent heat energy decline. The depression decays, or fills, as

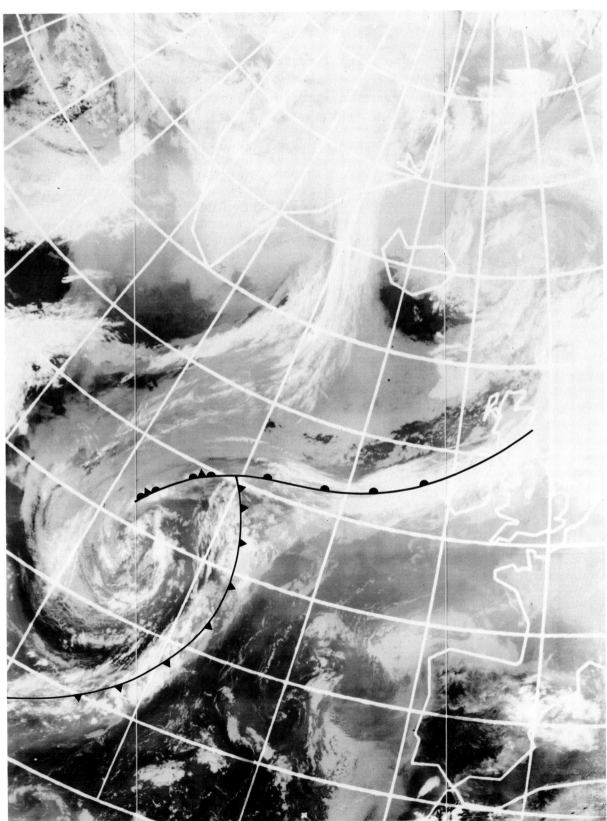

Plate 7.4   NOAA–5 infra-red 2116 GMT, 21 August 1978. (Photograph: University of Dundee)

pressure at its centre begins to rise. After a while, all that remains of the depression is a weakening cyclonic eddy within the maritime polar air mass (Plate 7.4). The polar front is then situated equatorwards of this eddy and along it the process of depression formation begins anew. During its life-cycle, therefore, a depression achieves an interchange of air between tropical and polar sources, a fact of considerable importance for the global energy balance mechanism described in Chapters 2 and 3.

2. Anticyclones

(a) Nature and characteristics

Anticyclones are atmospheric systems where relatively high air pressure is experienced at the surface, due to a downward and outward spiral of air from aloft. In the northern hemisphere, this takes the form of a clockwise circulation of air around the centre of subsidence, which, since it is fed from above, normally involves no surface air mass differences in the manner of a depression. Indeed, with their gentle breezes, absence of precipitation and often cloudless skies (due to the warming of the subsiding air), anticyclones are in many respects the antithesis of depressions. Stable and slow-moving, they are associated with warm, fine weather in summer and cold, frosty conditions in winter.

Figure 7.5   A polar outbreak in North America – surface weather chart (isobars given in mb; temperatures in °C) for 0000 GMT on 25 December 1983

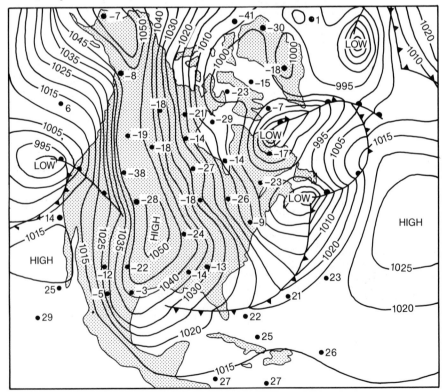

Plate 7.5 NOAA–5 infra-red 1911 GMT, 3 April 1978. (Photograph: University of Dundee)

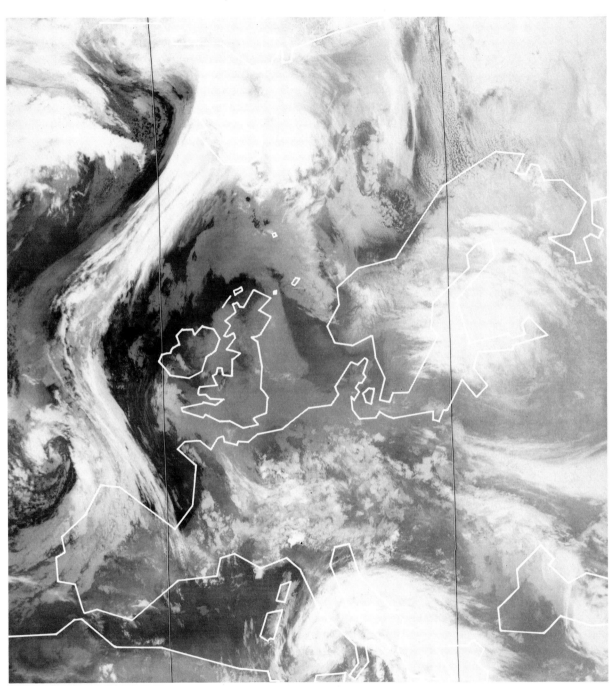

On some occasions, the anticyclonic activity affecting the mid-latitude zone is merely an extension of the semi-permanent high-pressure cells of the subtropics, particularly in summer, when these are displaced polewards. In winter, strong, though shallow anticyclones develop over the cold interiors of North America and Eurasia; when these intensify, bitterly cold spells may occur even as far south as the subtropical margin. Figure 7.5 shows such a polar outbreak in North America around Christmas 1983, when record low temperatures were experienced over many parts of the United States. More than 270 people were killed during a 10-day period in weather-related accidents, and considerable crop damage occurred as far south as Florida.

(b)   Blocking anticyclones

As described in Chapter 3, cellular fragmentation of the upper westerly flow may occur during times of low zonal index circulation (that is, large-amplitude waves). In the large cells of air detached from the mainstream of the circulation, the formation of quasi-stationary weather systems, including anticyclones, commonly occurs. Such anticyclones, once they become established, effectively block the west-to-east passage of weather. Often, during such a period of blocking, the upper westerly flow is split into two streams, bringing depression activity to areas far to the north and south of the main depression tracks (see Figure 6.20). Within the block, classic anticyclonic symptoms prevail, perhaps for several weeks. Thus both inside and outside the block persistent weather anomalies may be experienced over large areas.

Some locations are more prone to blocking than others, and seasonally preferred locations are often responsible for the recurrence of certain weather conditions at particular times of the year. For example, blocking anticyclones in the vicinity of Scandinavia in spring partially explain the reduced precipitation amounts over Britain and Ireland at this time of the year. The scenario for this is exemplified in Plate 7.5, which shows the operation of a blocking situation during spring 1978. The anticyclone is identifiable by the area of clear skies extending from the Baltic Sea to the west of Ireland. Depressions can be seen moving around the flanks of the anticyclone near Iceland, north-west of Portugal and in the Mediterranean Sea near Italy.

3.   Upper air and surface weather relationships

Fairly complex linkages exist between features of the upper air circulation and weather systems at the surface. To understand the nature of these requires a good grasp of the interaction between forces that produce air motion; these have been described in detail in section B of Chapter 3.

(a)   Convergence and divergence aloft

For depressions to deepen rapidly, as they frequently do, there has to be a mechanism for removing the rising air of the vortex even more rapidly

Figure 7.6 Convergence/divergence aloft and weather systems at the surface

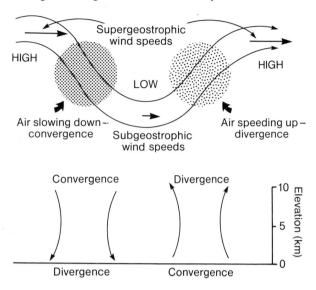

than air is supplied to it by convergence at the surface. This is obviously facilitated by the divergence of air at high levels in the troposphere. Accordingly, divergence aloft tends to be associated with depression formation (*cyclogenesis*) at the surface and vice versa, whereas convergence aloft encourages the downward motion of air, which may nourish an anticyclone at the surface (Figure 7.6). Clearly, the upper air environment exerts a controlling influence not only on the formation and intensity of surface weather systems, but also on their direction of movement. In these relationships, the upper westerly waves have an important role to play.

(b)   Curved airflow aloft – the gradient wind

For the examination of the geostrophic wind in Chapter 3, isobars were depicted as straight lines (see Figure 3.6). Frequently, however, isobars are curved, and air parcels aloft move in a curved path parallel to them. Such a flow of air in a curved path implies an imbalance between the Coriolis and pressure-gradient forces. Around a low-pressure centre, it implies that the pressure-gradient force is greater than the Coriolis force, and vice versa around a high-pressure area. Where such a flow runs parallel to the isobars, the wind is said to be a *gradient wind* to emphasise the absence of a balance between these two forces.

From Figure 7.7 it can be seen that the constantly changing direction of the pressure-gradient force along the curved isobars aids the build-up of wind speed around a high, whereas around a low the changing direction of the pressure-gradient force obstructs the build-up of wind speed. For a balance to be achieved between the pressure-gradient and Coriolis forces, therefore, a faster speed is required in highs than in lows. If the same pressure-gradient forces existed in the two systems, then it could be

Figure 7.7 Gradient wind around high and low-pressure systems. (Source: Trewartha and Horn, 1980)

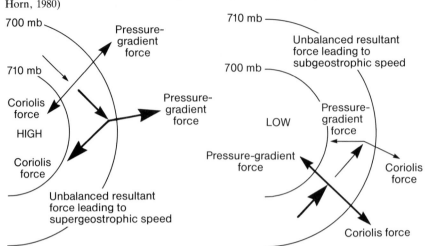

envisaged that the wind speed around high pressure should be greater than around lows.

In the ridges and troughs of the upper westerlies, air is thus moved through the ridges at speeds faster than the pressure-gradient wind (*super-geostrophic*), and through the troughs somewhat slower (*subgeostrophic*). These changes in speed produce a 'piling up' of air in the stretch between ridge and trough, where the forward air is slowing down as the trough is approached, whereas the air to the rear is still moving fast on leaving the previous ridge. This catching-up process is known as *convergence* and may force air downwards to the surface. In such circumstances, anticyclones may form at the surface, fed by the downward spiral of air from aloft. In the stretch between trough and ridge, on the other hand, the acceleration of air ahead of, in contrast to the slower-moving air emerging from the trough, produces a *divergence* of air. In this instance, air tends to be drawn up from the surface, perhaps giving birth to a depression in the process. Hence, there are areas under the upper westerly waves where depressions or anticyclones are encouraged to form –

Figure 7.8 Upper air motion and cyclogenesis

A Anticyclone development
C Depression development

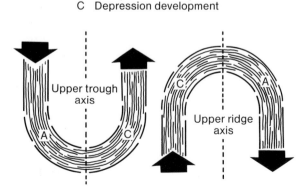

depressions on the poleward-running air approaching an upper air ridge, and anticyclones on the equatorward-running air approaching an upper air trough (Figure 7.8).

The most rapidly moving parts of the upper westerly circulation, represented by the jet streams, are particularly significant for depression formation. The location of the polar jet stream, as has been emphasised in Chapter 3, bears a close relationship to the polar front. The polar jet naturally provides an ideal removal mechanism for the rising air motions associated with the polar front, where the overwhelming bulk of depressions form. In places, the jet stream itself varies considerably in velocity. It is particularly fast-flowing on the extremities of its meanders. Downwind of these, the flow spreads out somewhat, producing the divergent conditions that encourage depression formation below.

ASSIGNMENT
1. *With reference to the case-study of a depression outlined above, describe and account for the sequence of weather changes that might accompany the passage of a mature mid-latitude depression from west to east, just north of your home area.*

C. Low-latitude Weather Systems

Climatologically, the tropics are best distinguished by their contrast with the temperate zones, and it would be reasonably accurate to describe them as being defined by temperature and differentiated by rainfall. This may be appreciated if the relative importance of these two weather elements in the mid and low latitudes is briefly considered.

In the mid-latitudes, the weather systems described in section B above bring precipitation, especially to the western margins of the continents. However, in the continental interiors, the smaller contribution to annual total precipitation from maritime depressions is compensated for somewhat by summer convectional rain. Overall, therefore, precipitation is not as effective in discriminating climatic regions as is temperature. Indeed it is to the latter that the rhythms of the living environment are predominantly attuned.

In the tropics, on the other hand, temperature is not the critical parameter for the most part. Seldom does the seasonal range exceed 10°C, and the diurnal variation is normally more than the annual range. The critical value of 0°C is never, or nearly never, approached; this has great significance for the proliferation of vegetation, insect life and indeed pathogenic organisms, which are absent in the mid-latitudes. Rather, it is rainfall that dictates the seasonal rhythm of life. For this reason, it is important to understand the nature of rainfall-producing weather systems embedded in the trade-wind circulation.

1. Disturbances in the trade-wind circulation

The most common type of tropical disturbance consists of westward-moving troughs of low pressure, normally extending polewards from the equatorial trough. These originate in the easterly trades aloft and develop mainly over ocean areas. In the case of the tropical Atlantic, the perturbation may frequently be traced back to disturbances over Africa, from which a westward-propagating wave emerges. Enhancement to wavelengths typically between 2000 and 3000 km is facilitated by the fact that the disturbance is embedded in the trade-wind circulation. At low levels, these winds blow from north-east to south-west in the northern hemisphere, meaning that air is blowing over progressively warmer and warmer areas of ocean. This causes the moist air close to the surface to become heated and great instability to develop. Cumulus towers punch upwards into the stable air aloft and the disturbance becomes organised into a cloud system which often shows a distinctive 'inverted V' configuration. About fifty such waves cross the tropical Atlantic each year from east to west, at speeds of 400–800 km/day.

2. Cyclonic disturbances

(a) Hurricanes: nature and distribution

About 10% of wave disturbances intensify into the more violent rotating storms so feared in many parts of the tropics. Known by a variety of names – hurricanes (Atlantic and eastern Pacific Oceans), typhoons (western Pacific Ocean), cyclones (Indian Ocean), baguios (Philippines), willy-willies (Australia) – these disturbances cause more fatalities on average than any other natural disaster, with the possible exception of drought. They have been intensively studied, particularly by satellite, as equipment failure is common in wind speeds which often greatly exceed 50 m/s.

Figure 7.9 shows that hurricanes are maritime phenomena, originating over tropical oceans where sea-surface temperatures are in excess of 27°C. Over colder ocean areas, and particularly over land, they seem to dissipate rapidly. A constant supply of warm, humid air thus appears to be a primary nutrient for hurricanes. A second prerequisite relates to latitude. Figure 7.9 also shows a hurricane-free belt close to the equator, where formation is inhibited. This certainly relates to the very weak Cor-

Figure 7.9   Global distribution of hurricanes with respect to sea-surface temperature (in °C)

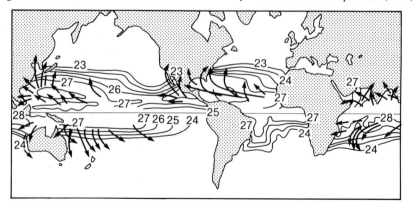

162

iolis force existing at these latitudes. Since the tendency for displaced air to rotate increases polewards, hurricane formation evidently requires a modicum of 'spin' forces to trigger off rotation. These two necessities explain why hurricanes form usually at 5–8° N and S, over the tropical oceans. In addition, it is in late summer that the maximum poleward extension of the required warm ocean waters occurs and so the months July–October constitute the main hurricane season in many areas of the northern hemisphere.

(b) Mechanisms of hurricane formation

(i) *Youth*. The mechanism of hurricane formation appears, therefore, to be as follows. First, for some small-scale initial disturbance, warm, humid oceanic air is induced to rise. If sufficiently vigorous, and if the location is far enough away from the equator, the effects of the rotation of the earth give the rising air a 'twist' and the whole system begins to revolve. The organisation of clouds into spiral bands is a critical stage in this process, transferring energy from individual rising packages of air into a more coherent vortex. Simultaneously, large amounts of latent heat energy are released through condensation, further enhancing instability, drawing in more humid, oceanic air from adjacent areas, and accentuating the scale of the system (see Figure 7.10). Mature hurricanes range in diameter: 200–500 km is typical. Central pressure is usually below 970 mb, exceptionally below 880 mb, and around this vortex the full fury of the storm is experienced. A towering wall of cumulonimbus clouds reaching 12 km high is evidence of the intense convection activity fuelled by latent heat release. Up to 15×10^9 tonnes/day of water vapour passes through the system, half of which may fall as rain. Flooding, structural damage due to winds sometimes over 250 km/h, coastal inundation due to storm surges from the ocean: these explain the havoc which accompanies a hurricane (see Plate 7.6).

Figure 7.10   Structure of a mature tropical hurricane. (Source: Brimblecombe, 1981)

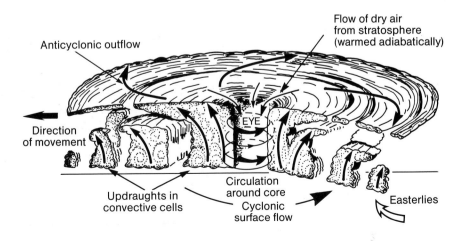

Plate 7.6 Effects of Hurricane Hazel. In autumn 1954, Hurricane Hazel caused great devastation in Toronto, Canada. This is all that remained of one home, after the combined effects of high winds and flooding (Photograph: Toronto Star Syndicate)

(ii) *Maturity: the eye of the storm*. At the heart of the cloud spiral is often a quiescent area, perhaps 30–50 km in diameter, with clear skies and light winds. This is the '*eye*', a small area where subsidence of air from aloft occurs. Its importance is twofold. First, it signifies divergence aloft, a vital aspect in removing the rapidly rising air of the hurricane and permitting convection to proceed at the surface. Second, the warming of the descending air in the eye itself induces instability at the surface and stimulates the storm's intensity. These features of divergence aloft and convergence at the surface mean that the internal structure of a hurricane resembles that shown in Figure 7.10.

(iii) *Decay*. The system is self-perpetuating as long as a supply of warm, moist air nourishes it. Movement over a warm oceanic area at about 20 km/h is ideal. Landfall, or passage over colder ocean, destroys the disturbance, although on occasion it may become absorbed into the mid-latitude westerly circulation and recurve eastwards as an intense mid-latitude depression. An unusual chart is depicted in Figure 7.11, showing four hurricanes simultaneously in the Atlantic/Gulf of Mexico area. The typical recurving track can be seen. Debbie, in particular, eventually reached Iceland with still hurricane-force intensity, having caused 17 deaths in Ireland.

Figure 7.11 Isobaric chart (in mb) showing four hurricanes in the Atlantic/Caribbean area. (Source: Riley and Spolton, 1981)

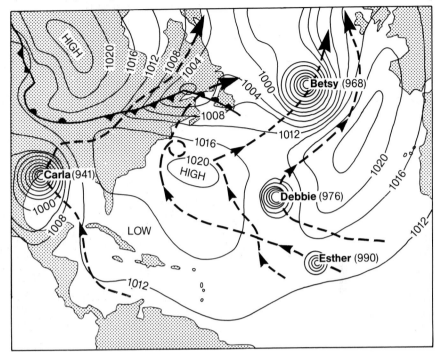

(c) Hurricanes: a case-study

A more typical progression can be seen for Hurricane Gloria in September 1979 (Plates 7.7–7.9). In Plate 7.7, the tell-tale comma cloud formation of an incipient hurricane can be seen off West Africa. Nine days later (Plate 7.8), Gloria has matured into an intense rotating disturbance in the western Atlantic. The spiral cloud configuration and the eye can be clearly seen. This, however, was her finest hour and only one day later Gloria was in decline. In Plate 7.9, her incorporation into a mid-latitude disturbance was clearly underway. The tight cloud spiral was disintegrating into frontal-like cloud bands and the hurricane stage was over.

Given their potentially destructive effects, attempts have been made to divert or dissipate hurricanes before they reach land by seeding them with silver iodide crystals. It is a reflection of human ingenuity that a few hundred kilograms of chemicals may influence a system with an energy equivalent of eighty 100-megatonne nuclear bombs per day. However, such experiments are still inconclusive and raise ethical and political as well as scientific problems, since many tropical areas depend on rainfall from hurricanes for agriculture. Hurricanes are also a vital component of the global heat transfer mechanism, redistributing energy polewards. For these reasons, attempts to interfere with these tropical disturbances must proceed with caution.

Plate 7.8   Meteosat visible 1155 GMT, 13 September 1979. (METEOSAT image supplied
by the European Space Agency)

Plate 7.9   Meteosat visible 1155 GMT, 14 September 1979. (METEOSAT image supplied
by the European Space Agency)

1. (a) Outline the factors conducive to hurricane formation.
    (b) Examine Figure 7.9 and suggest reasons why no hurricane de-
         velopment occurs in the tropical Atlantic south of the equator.
    (c) Suggest why the wind speeds in hurricanes decrease rapidly when
         they cross land, although heavy precipitation may continue for
         some time.

2. The following is the text of a hurricane warning issued by the Miami
    Weather Bureau:

    > Dangerous hurricane Cleo continues to move toward the south-east
    > Florida coast and safety precautions against hurricane winds, heavy
    > rain and above normal tides should begin immediately and be
    > rushed to completion with all possible urgency in the area of
    > hurricane warning display from West Palm Beach to Key Largo.
    >
    > At 2 p.m. the center of the well-organized and intensifying
    > hurricane Cleo was centered by land-based radar and air recon-
    > naissance about 200 km south-south-east of Miami. Cleo is moving
    > toward the north-west at 27 km/h and is expected to be just a very
    > short distance south-south-east of Miami early this evening.
    >
    > Highest winds are estimated to 175–200 km/h near the center.
    > Hurricane force winds will extend 80 km from the center and gales
    > around 160 km in all directions.
    >
    > Hurricane Cleo is expected to skirt the Florida coast tonight.
    > However, a slightly more north-westerly course would mean a direct
    > hit on the coast. This would mean that storm tides of possibly 2.5 m
    > would develop near and just north of the center.

    (a) Plot on a sketch map the present and predicted location of hur-
         ricane Cleo. Outline the areas affected by (i) hurricane winds and
         (ii) gales.
    (b) Suggest why the storm tide will be highest just north of the eye.
    (c) Make a list of safety precautions that a resident of Key Largo
         should take.

Key Ideas

A. Air Masses and Fronts

1. A body of air with relatively uniform horizontal distribution of tem-
    perature and pressure is known as an air mass.
2. Air masses develop when air stagnates over places with homogeneous
    surface conditions.
3. Areas over which air masses form are called source regions.
4. Air masses can be classified on the basis of temperature condition and
    moisture content.
5. Fronts are zones of air mass interaction and frequently involve uplift,
    linear cloud developments and precipitation.

B. Disturbances in the Mid-latitude Circulation

1. Convergence of contrasting air masses along the polar front prompts the development of cyclonic disturbances known as depressions.
2. Depressions generally move from south-west to north-east, controlled by the westerly flow aloft, and bring a distinctive sequence of weather types as their associated frontal zones cross an area.
3. Depression tracks may be displaced during periods of slack or meandering upper air flow and blocking anticyclones develop, which bring anomalous weather types over large areas.
4. Convergence in the westerly flow aloft between ridges and troughs encourages anticyclone formation at the surface.
5. Divergence in the westerly flow aloft between troughs and ridges encourages depression formation at the surface.

C. Low-latitude Weather Systems

1. No marked air mass differences, no marked thermal gradients, no marked seasonal temperature contrasts exist in the tropics.
2. Linear disturbances resulting from low-pressure troughs moving westwards in the trade-wind circulation are the most common cause of precipitation on eastern continental margins in many parts of the tropics.
3. Very intense rotating storms develop occasionally during the summer over areas of very warm ocean. Once initiated, these are powered by release of latent heat through condensation and move westwards, intensifying into large-scale cyclonic vortices.
4. Hurricane formation is strongly influenced by upper air considerations. High pressure aloft allows divergence to remove the rapidly rising air from below and permits warming due to subsidence in the eye of the hurricane.
5. Hurricanes decay when their source of warm, moist maritime air is diminished, although they may metamorphose into intense mid-latitude depressions on occasion.

Additional Activity

1. Study Plate 7.10, which is an infra-red image of Western Europe taken by the satellite NOAA–7 at 1507 GMT on 20 February 1984. Colder and hence higher clouds appear brighter on this image than the greyer tones of their warmer, lower counterparts. Note, in particular, that the spiral of cloud associated with a depression, with central pressure of about 972 mb, is located just off western Ireland. Relatively clear skies further east are evidence of an anticyclone centred over northern Sweden, where a pressure of 1040 mb was observed at this time.
    (a) Construct a sketch map which classifies the areas of cloud into (i) high stratiform clouds, (ii) low stratiform clouds and (iii) cumulus.

Plate 7.10 NOAA–7 infra-red 1507 GMT, 20 February 1984. (Photograph: University of Dundee)

(b) Construct a simple isobaric chart which might represent the situation imaged, with isobars at 4 mb intervals. Mark on it the position of the warm and cold fronts associated with the depression.

(c) Suggest reasons for (i) the cumulus development in the central Atlantic and (ii) the low stratus forming over the North Sea.

(d) Prepare a weather forecast, as detailed as you can make it, for London for the next 24 hours.

8 Climatic Change

A. Change as the Norm

Variability over a great range of time and distance scales is an important characteristic of the atmospheric system. When aggregated and averaged over many years a unique climatic fingerprint emerges, expressed in terms of means, extremes and frequencies of various weather elements. Conventionally, 30 years of observations (for example, 1901–30 or 1931–60) have been used as standard reference periods to establish the so-called climatic norms of a place. Such an exercise had many advantages, not least of which was the opportunity it afforded of making statistical estimates of the likelihood of specific departures from the norm occurring over time. Thus, for example, building engineers could be provided with estimates of the greatest gust of wind, or water engineers with the greatest intensity of rainfall, or insurance companies with the likely incidence of storm or frost damage over 50, or even 500 years, all extrapolated from a 30-year span of observations.

Shortcomings in this approach to climatic appraisal have become obvious in recent years. It is now clear that such 'baselines', established on the strength of only three decades of observation, cannot be extended either into the past or, more seriously, into the future with any degree of certainty. The norm for one period may well be exceptional when viewed in a longer perspective of time. For example, Figure 2.15 implies that the present century is warmer than the last century on average. In fact the interval 1901–60 may have been the warmest and wettest such period globally for almost a millennium. The present norm is instead the exception, and in a world geared towards a continuation of present climatic 'norms', the existence of continual and significant climatic variability has profound repercussions. The following quotation epitomises the issues:

> The climate of the earth is now known beyond any doubt to be in a more or less continual state of flux. Changeability is an evident characteristic of climate on all reasonable time scales of variation, from that of aeons down to those of millennia and centuries. The lesson of history seems to be that climatic variability is to be recognised and dealt with as a fundamental quality of climate, and that it should be potentially perilous for man to assume that the climate of future decades and centuries will be free of similar variability. (Mitchell *et al.*, 1975)

Climatic events during the 1970s and 1980s, such as the Sahelian drought described in Chapter 6, have demonstrated the truth of this statement. This and other climatic shocks have underlined the fact that, despite technological advances – perhaps even because of them – humankind is still vulnerable to the vagaries of climate. This realisation has prompted the development of many techniques that enable past climatic fluctuations to be investigated and possible causes hypothesised.

ASSIGNMENTS
1. *The following are the chances in 100 of frost not occurring after the specified date for part of the grain-growing belt of the mid-western USA:*
   10 April 1.5   20 April 10.0   30 April 33.0
   10 May 65.0   20 May 90.0   30 May 98.5
   Graph these values (if possible on normal probability paper) and hence estimate the date on which a farmer should sow in order to have a 50 : 50 chance of avoiding frost damage.
2. *What information would it be desirable to have, before relying on the above procedure?*

B. Reconstructing Past Climates

1. Documentary sources

Although the barometer and thermometer were invented in Italy in the seventeenth century, outside Europe instrumental records often extend back only for about a century; thus the chronological scope of this most satisfactory type of documentary evidence is very limited. Within Europe, regular observations commenced earlier, with isolated records dating from the mid-seventeenth century. The main impetus for the growth of the observational network, however, came as the industrial revolution gained momentum. In the rapidly expanding cities of the coalfields, the growing demand for public-water supplies necessitated the collection of weather data.

(a) Early direct sources: problems of standardisation

In seeking to use these early records for climatic reconstruction, caution must be exercised. A miscellany of instruments with different accuracies, non-uniform exposures, using a multiplicity of scales, means that painstaking analysis and comparisons of overlapping records is necessary before a reliable series can be derived. The longest such series for temperature is for a typical lowland site in central England, extending from 1659 (Figure 8.1). Fluctuations, even over this relatively short snippet of time, of up to 2°C in annual and seasonal averages are apparent. Even these are capable of inducing much greater changes in other significant climatic aspects, as will be demonstrated later.

Many other documentary sources exist, such as those of private individuals who have left a legacy of diaries and journals. These provide

Figure 8.1 Seasonal and annual temperatures for a typical lowland location in central England: ten-year running means, from 1659. (Source: Lamb, 1982)

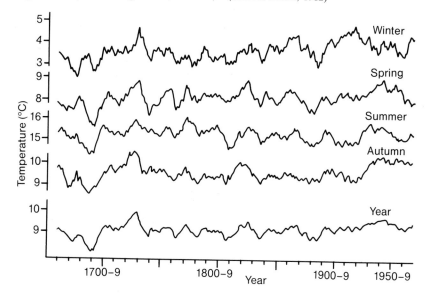

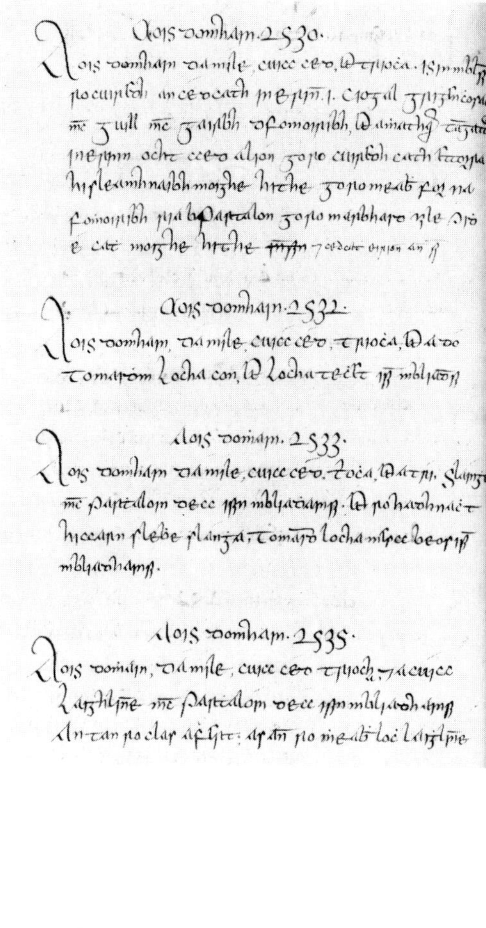

Plate 8.1 A storm on Lough Conn in 4668 BP. In this ancient Irish manuscript, reference is made to a storm on Lough Conn, in County Mayo, which is alleged to have occurred more than two millennia before the birth of Christ. Although such sources are partly legend and partly truth, they can occasionally enable inferences on past climatic conditions to be made. (Photograph: Royal Irish Academy)

weather information for individual seasons and years as far back as the fourteenth century, and may be supplemented by ships' logs, private estate records, early newspapers, personal writings, etc. Prior to about AD 1100, however, the documentary record is very fragmentary, consisting mostly of reports of notable extremes of weather, contained within various annals and chronicles. A long tradition of these exists in some countries such as China, Iceland and Ireland (Plate 8.1), enabling inklings of conditions in the pre-Christian era to be gleaned.

(b)   Indirect sources: a more objective measure?

Apart from direct references, past climatic conditions may often be inferred from an examination of their effects. Such 'proxy' material includes crop yields, grain prices, glacier movements and harvest dates. Often these may be more meticulously documented than weather conditions *per se*. For example, the blossoming of certain flowers or the ripening of fruit is mainly a function of the temperature to which they have been exposed since budding. The warmer and sunnier the vegetative period has been, the swifter the bloom or fruit reaches maturity. In the vineyards of northern France, the grapes are adjudged mature and the harvest date fixed by a panel of experienced growers in each area, a system in operation from the fourteenth century to the present day. Recent parts of this documentary record can be calibrated against instrumental records, and conditions in the pre-instrumental period inferred. A similar phenological record exists in Japan, where the emperor travelled each spring to the ancient capital of Kyoto to view the new cherry blossoms.

2.   Tree-ring analysis

(a)   Theoretical considerations

One of the most valuable indirect techniques allowing climatic reconstruction lies in the investigation of tree rings. Many trees produce an annual growth layer which appears as the outermost of a series of rings visible in a cross-section through the trunk. This is because each ring usually is formed from a cycle of growth that begins vigorously in spring and ends sluggishly in the autumn. The ring boundary is the abrupt change between the small, thick-walled cells of slow autumn growth and the large, thin-walled cells of the growth spurt occurring the following spring. If still growing, the approximate age of the tree may be ascertained by counting the annual rings.

Variations in ring width from year to year are often apparent, indicating the presence, or absence, of growth-limiting factors. At a local scale this may be associated with insect damage or management practices (changes in root competition, shade, etc.), but normally similar ring patterns are observed over large areas. The only conceivable agent with a sufficiently widespread and consistent influence on tree growth is climate, and this means that tree rings are a potentially valuable source of proxy climatic records.

(b) Methodology

The procedure involved for climatic reconstruction using tree rings (*dendroclimatology*) entails, first, the collection of a number of samples using a borer. The cores obtained are then examined and distinctive runs of ring widths used to establish where one record overlaps another. Perhaps some cores were obtained from stumps of old trees, old timbers or tree trunks buried in peat; but given a sufficiently large sample, all may be tied in to one sequence. This is the principle known as *cross-dating* (Figure 8.2) and enables each ring to be dated. Once all the rings are dated and measured, each core is standardised to eliminate the gradual changes in ring width occurring as the tree ages. A common index now enables the various samples to be merged into a single chronology which may span several centuries. One such chronology relevant to the British Isles commencing in AD 401 is shown in Figure 8.3.

A variety of statistical techniques is employed to extract the climatic information from the ring-width indices. Variability in these may occur for different reasons. Close to the polar or altitudinal tree line, tempera-

Figure 8.2   Cross-dating tree rings from various sources. (Source: Fritts, 1976)

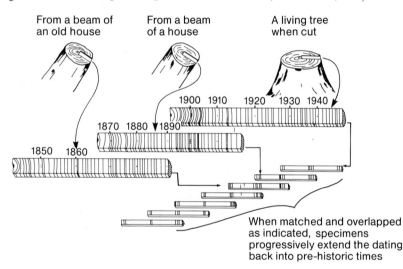

Figure 8.3   An oak ring-width chronology for the British Isles. (Source: Pilcher and Baillie, personal communication)

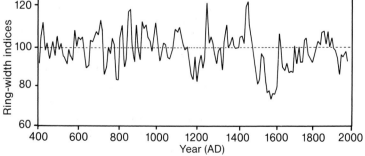

ture may be the critical factor limiting growth, whereas on arid margins precipitation may be most important. A knowledge of the precise nature of these relationships may be used to reconstruct climate or climatic anomalies in the more distant past, when instrumental records were not available.

3. Pollen analysis

Many plant species have particular climatic requirements for viability which condition their geographical distribution. Assuming that such requirements do not change with time means that the reconstruction of past vegetation distributions may provide a proxy record of past climates. Pollen grains are frequently used as indicators of previous vegetation distribution and thus as tools aiding climatic reconstruction. They possess the following characteristics.

(i) *Profusion.* Pollen grains are produced in great abundance by many species (for example, a single rye plant may yield 23 million grains), and are thus available in most environments for analytical purposes.

(ii) *Resistance to decomposition.* Nature protects the pollen grain by enclosing it in an outer wall composed of a highly resistant organic substance; if they come to rest in an oxygen-free environment, the grains may resist decay almost indefinitely.

(iii) *Distinctive nature.* The design of the outer wall of a pollen grain is distinctively different from species to species; thus it is feasible to identify from which species an individual pollen grain originates.

(iv) *Ability to be counted.* Because of this ability to be counted, the proportion contributed by various species to a pollen sample may be calculated. The composition of a pollen sample can accordingly be related to the composition of the contributing vegetational sources during the time at which the pollen grains were deposited. The major assumption of the technique is that the sample is reasonably representative of conditions in the surrounding area when allowance is made for different pollen productivities, different frequencies of flowering, different dispersal mechanisms, etc. between species. This assumption must, however, be carefully borne in mind at all times.

(v) *Ability to be preserved in layers.* In some areas, such as lake floors and peat bogs, each year's pollen 'rain' is deposited on top of the previous year's in an undisturbed succession, which enables changes over time in vegetation composition, and by implication climate, to be examined.

(vi) *Possibility of dating pollen layers.* Modern dating techniques, such as radio-carbon dating, allow absolute dates to be added to pollen records.

In Figure 8.4, the fluctuations of recent centuries, previously seen in Figure 8.1, can now be seen to be relatively minor on this longer-time perspective of 20 000 years. Rather, a range of variation of greater than 10°C

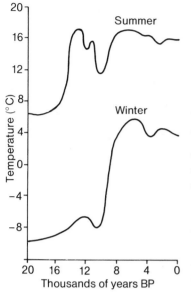

Figure 8.4 Temperature trends in Britain as derived from pollen analysis. (Source: Lamb, 1982)

can now be detected with a relatively rapid change-over between two climatic modes – glacial and interglacial. The rapid emergence from glacial conditions around 10 000 BP (before present) is particularly striking in the steep rise of winter temperatures. The warmest post-glacial times are also detectable (8000–4000 BP) with a gentle, irregular decline in both summer and winter temperatures occurring since then to the present day.

4. Isotope analysis

(a) Principles of isotope analysis

Variation in the number of neutrons in the atomic nucleus of a particular element gives rise to isotopes of that element. In nature, most elements occur as a mixture of such isotopes; for example, the natural abundance of each of the three isotopes of oxygen are as follows: ^{16}O 99.759%, ^{17}O 0.037% and ^{18}O 0.2039%. Although the chemical and physical properties of the various isotopes are similar to those of the parent element, there are often slight differences arising from their differing nuclear masses. It is these slight differences which provide a tool for unravelling the temperature history of the earth for over one million years before the present.

In climatic studies most attention has been given to the isotopes of oxygen, because that element exists as part of so many compounds, especially in the large reservoir of the oceans. Incorporated in water, water vapour or ice, oxygen is a part participant in the great recycling mechanism of the hydrological cycle (see Chapter 4). The slightly different behaviour of the two isotopes ^{16}O and ^{18}O as they move through this cycle provides the basis of isotope studies.

When water evaporates, such as from the ocean surface, the small difference in mass means that ^{16}O-based water molecules enter the vapour phase slightly more readily than their heavier ^{18}O counterparts. The remaining water thus becomes enriched in ^{18}O and relatively depleted in ^{16}O. Condensation reverses the process and restores the equilibrium. If, however, large continental ice masses exist, this return flow is diminished as precipitation becomes locked up as ice. The more extensive the ice-sheets are, the heavier the isotopic composition of the sea water becomes. Measurement of this isotopic enrichment of heavy oxygen in sea water, or of its depletion in precipitation, is thus an indicator of the extent of global ice masses and, by inference, of temperature.

(b) Ice core investigations

To use this finding in determining past temperatures necessitates discovering deposits which have recorded these variations of isotope abundance in datable strata. One such source is the polar ice-sheet, where the accumulation of snowfall in annual layers of ice occurs. A core from a region with little summer melting provides a record of precipitation extending back in the form of easily counted annual layers. The isotopic composition of each layer of ice is a record of the isotopic composition

177

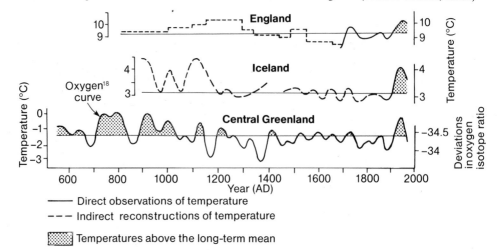

Figure 8.5   Temperature trends in Greenland derived from ice core analysis compared with temperatures measured and inferred for Iceland and England. (Source: Gribbin, 1978a)

— Direct observations of temperature

--- Indirect reconstructions of temperature

Temperatures above the long-term mean

of the precipitation which produced it. Where such a layer is depleted of heavy oxygen isotopes, the temperature conditions under which the precipitation occurred must have been cold and vice versa. A proxy record of temperature exists. Some very deep cores have now been driven into the polar ice-caps. One extends to over 2 km in Antarctica and provides data going back 75 000 years. Figure 8.5 shows a comparison between the isotope record for a core in central Greenland and temperatures measured and inferred for Iceland and England back to the ninth century. A convincing correlation between the isotope record and the directly observed data is apparent.

(c)   Ocean core investigations

Sediments cored from the ocean floor also provide a record. Plankton, tiny temperature-sensitive lifeforms, exist in huge numbers in the ocean and, as they die, settle in layers on the ocean bed. This constant rain of tiny bodies accumulates at rates of 1 or 2 mm per century. Some types are only found in waters warmer than 10°C. Others have a spiral shell which coils to the left when the prevailing water temperature is colder than 7°C and to the right when it is warmer. Mapping the distributions of such indicator species enables water temperatures for the various layers to be reconstructed.

More recently, isotope analysis of the shells of certain plankton (for example, Foraminifera species) has provided a powerful tool with which to reconstruct past atmospheric temperatures and the volume of past ice-sheets. Examinations of the ratio of ^{16}O and ^{18}O in the shells indicates whether the ocean water used in their construction was depleted in ^{18}O. Since this is a surrogate index for temperature, such an analysis yields the possibility of establishing climatic conditions back into the early part of the Pleistocene epoch. Figure 8.6 shows one reconstruction of conditions during the peak of the last ice age.

Figure 8.6 Climatic reconstruction for 18 000 BP: (*a*) July sea-level pressure values (mb) and (*b*) July temperaturs (°C). (Source: Lockwood, 1979)

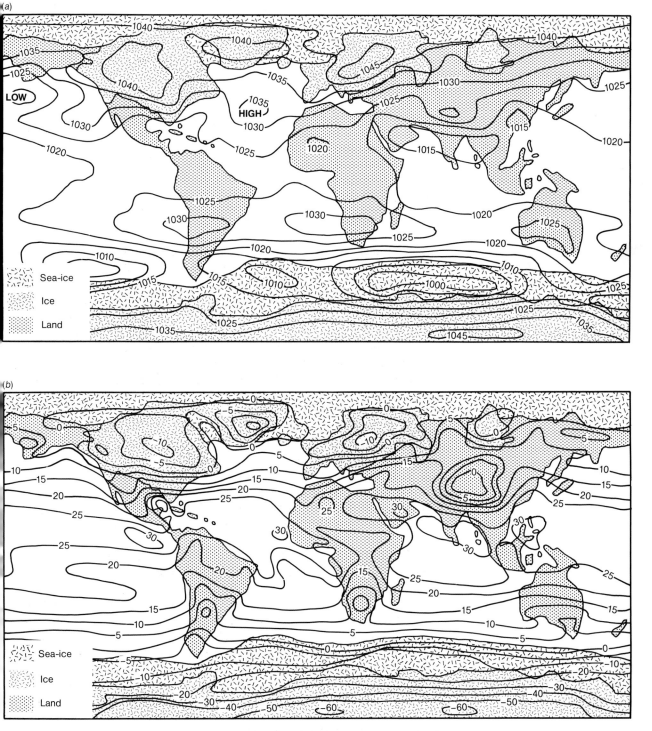

ASSIGNMENTS

1. (a) *Construct a histogram from the following documentary record showing the number of weeks per year (AD) on average when polar ice was recorded close to the coast of Iceland:*

900–60	0	960–80	2	980–1000	0	1000–20	1
1020–1200	0	1200–20	1	1220–40	3	1240–60	2
1260–80	4	1280–1300	2	1300–20	4	1320–40	2
1340–60	0	1360–80	4	1380–1460	0	1460–80	4
1480–1500	0	1500–20	1	1520–40	0	1540–60	2
1560–80	4	1580–1600	2	1600–20	13	1620–40	16
1640–60	4	1660–80	6	1680–1700	13	1700–20	10
1720–40	5	1740–60	18	1760–80	8	1780–1800	26
1800–40	18	1840–60	9	1860–80	21	1880–1900	18
1900–20	10	1920–40	2	1940–60	3	1960–80	9

(b) *Describe the climatic trends which might be inferred from AD 900 onwards.*

2. *For each of the following techniques (a) outline the principles which enable them to be used in deriving past temperature trends and (b) suggest weaknesses which they may possess for this purpose: wine harvest dates, pollen analysis, tree-ring analysis, isotope analysis of ice cores.*

C. The Climatic Record

1. Glacial climates

Throughout its geological history the earth has been prone to climatic changes sufficiently dramatic to induce the rapid growth of glaciers at high latitudes and their expansion as continental ice-sheets covering up to one-third of the land area of the globe. Such periods of glacial expansion, lasting around 100 000 years, are then abruptly terminated by a climatic oscillation to relatively warm, interglacial conditions. The latter climatic mode, which prevails at present, typically persists for about 10 000 years before a reversion to glacial conditions ensues. Many such oscillations are now known to have occurred during the present series of glaciations, which commenced about 2 million years ago.

The last major glacial expansion reached its maximum extent about 18 000 BP, when an area exceeding 40 million km^2 was covered by ice up to 3 km thick. This enormous area of ice would obviously have considerably modified the general circulation of the atmosphere, compressing circulation types and climatic zones towards the equator. This may be simulated using the computer models of the circulation described in section D of Chapter 3. When run with data inputs appropriate to ice-age earth, these give results which agree reasonably well with those suggested by other climatic reconstruction techniques. Figure 8.6(a) depicts simulated July sea-level pressure values for approximately 18 000 BP. The most significant departure from the present is clearly the existence of prominent anticyclones over each of the major continental ice-sheets. This is a response to the chilling of the layers of air close to the surface in these ice-covered areas. In Europe and North America, the cold north-easterly air stream spiralling out from these anticyclones is responsible for the temperatures around 10°C below those of today over large areas south of the ice-sheets (Figure 8.6(*b*)). Sea-surface temperatures do not appear to have been greatly affected overall, although poleward-moving warm currents are obviously greatly restricted in their penetration into higher latitudes. The North Atlantic Drift, for example, is severely hindered by the presence of sea-ice and the prevailing north-easterly winds.

The rapidly changing climatic milieu of the last 2 million years was of considerable significance for the distribution and evolution of plant and animal life. Vegetation zones and habitats ebbed and flowed with the climatic alternations. In such circumstances, those animals who could adapt best to changed conditions – mobile, omniverous, with greater brain power – were selectively favoured. Foremost among these were humans, whose ecological dominance has its roots in this period. Riding the wave of glacial/interglacial oscillation, this species expanded out of the low latitudes to tackle new habitats laid bare by the retreating ice. Although adaption was facilitated by the cultural devices of fire, clothing, shelter and tools, the chronological coincidence of important environmental and cultural change during the later stages of the period of glacial/interglacial climates cannot be ignored.

2.   Course of the post-glacial climate

(a)   *Post-glacial warming*

Rapid warming followed the ending of the last glacial episode about 10 000 BP. In the northern hemisphere, this was accompanied by a northward shift of vegetation zones, as species formerly banished to southerly locations recolonised the fresh boulder-clay soils to the north. Sea levels rose rapidly, at about 1 m per century, submerging many coastal plains. Some of these, like the broad plain which became the North Sea, were inhabited, and their inundation, achieved by sporadic storm surges, probably cost many lives. By 7000 BP the islands of Britain, Ireland and Denmark were detached from continental Europe, bringing virtually to an end the natural immigration of most plant and animal species and explaining the floristic and faunistic impoverishment of these islands today.

(b)   *Post-glacial optimum*

The remnant of the Laurentian ice-sheet in Canada was slower to disappear than its European counterpart, and the presence of significant amounts of ice here as late as 5000 BP had important climatic implications. Figure 8.7 shows a reconstruction of the circumpolar vortex for a typical summer *c.*8500 BP. The enhanced trough over Labrador is a response to the ice-covered surface and would have had the effect of steering depressions to the north of a Europe dominated by tropical air masses.

The poleward displacement of climate zones was not confined to Eur-

Figure 8.7 Reconstruction of the circumpolar vortex, approximately 8500 BP (mean height in metres of the 500 mb surface in July) (Source: Lamb, 1982)

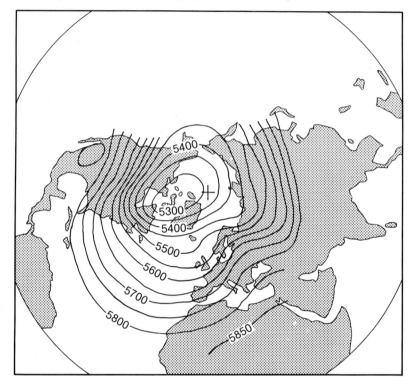

ope. In Africa much moister conditions existed in the Sahara, where rock drawings depicted an agricultural economy based on cattle herding and a landscape with elephants, hippopotamuses and giraffes in places where rain is seldom experienced today. In Egypt this was the period of the Early Kingdom cultures, and evidence suggests that the discharge of the Nile was treble what it is now.

For much of the world, the warmest climates of post-glacial times were experienced during the period 7000–5000 BP. In the mid-latitudes, summer temperatures were 2–3°C higher than today.

(c) Climatic record 5000–2000 BP

Cooling, and a return to drier conditions, became established after 5000 BP. Increased variability of rainfall disadvantaged many regions. Concentration of animals and humans around reliable sources of water, such as oases or river valleys, occurred. It has been suggested that the riverine civilisations of Mesopotamia, the Nile, Huang-He (Hwang-Ho), and Indus offered irrigation potential as a reward for social organisation, and the refugees from outside provided the slave labour necessary to realise this potential. Further north, a return to moister conditions is apparent between 3000 and 2000 BP. At Tregarron, in Wales, the bog thickness increased by as much during 400 years of this period as it did during the following 2000 years. Storms and rain appear to indicate a period of strong westerly winds, with the polar front close to the northern shores.

3. Medieval warm period

The cultural stagnation of much of Europe following the demise of the Roman Empire did not extend to its maritime fringes. Extensive voyages of discovery were carried out by Irish monks, who sought to establish monastic footholds in lands which the annual migrations of the wild geese suggested lay to the north and west. The annals record visits to Iceland and to the edge of the polar sea-ice from the sixth to the ninth century AD. Thereafter, the Vikings become prominent. In 865, an unsuccessful attempt by them to colonise Iceland led to the naming of that island as such in the sagas. Within a decade, however, a rapid climatic warming was underway, one which was experienced throughout much of the world, and which enabled the Vikings not just to colonise Iceland, but to export people from there to another promising island, Greenland. The warming of the ninth century may thus partly explain why Greenland was not called Iceland, and vice versa, something which would certainly be more logical today.

On the European mainland, cereals were being grown at higher altitudes and latitudes than ever before. Traces of ploughed fields still exist on the Scottish borders up to 320 m, where such agriculture today would be hopelessly unviable. Many vineyards were established in England, some as far north as York, indicative of summer temperatures 1°C or so warmer than today. At times temperatures approached those of the warmest centuries during the post-glacial record, and the period is often referred to as the Little Optimum. By the thirteenth century, however, a downturn was apparent, and by 1342 the traditional sailing route between Iceland and Greenland had to be abandoned in favour of a more southerly one that was less affected by the encroaching ice. The incidence of storms, disease and harvest failures increased sharply in Europe, where famine, and even cannibalism, was reported following the failure of the grain to ripen in 1315. A prolonged deterioration of climate followed, although it was punctuated by several periods of amelioration.

4. Little Ice Age

Figure 8.8 Little Ice Age temperature trend. (Source: Gaskell and Morris, 1979).

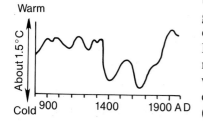

Harvests have always been a matter of life and death for the majority of the world's population. A relatively small deterioration of climate may thus be crucial in areas where wresting a living from the soil is only marginally feasible (see Plate 8.2). Such was the experience of many areas during what was probably the coldest climatic snap since the retreat of the Pleistocene glaciers. Termed the Little Ice Age, this episode encompasses much of the interval from the thirteenth to the mid-nineteenth century, with the most severe phases occurring between the sixteenth and eighteenth centuries. A global cooling of the order of 1–1.5°C was involved (Figure 8.8).

183

Plate 8.2 Drying the hay on lines in the Upper Setesdal valley, Norway. The difficulties of agriculture in a climatically marginal environment are illustrated here, where the warmth of summer is often insufficient to dry the hay properly and it is hung on lines to maximise the drying power of sun and wind. (Photograph: J. Sweeney)

(a) Little Ice Age in Greenland

One of the most marginal areas was in Greenland, where the Norse settlements by the twelfth century totalled about 300 farms. Supplying this outpost became increasingly difficult, and regular communications ceased after 1369. One ship, blown off-course in 1540, found no inhabitants alive. Excavations of the graveyards more recently paint a chilling picture of the colony's extinction. In the older graves the men were on average 178 cm (5 ft 10 in) tall. By the fifteenth century the average male height had declined to 165 cm (5 ft 5 in). Furthermore, the older graves were in deep trenches, whereas the more recent ones were progressively nearer the surface; the encroaching permafrost was obviously responsible. Today the surface is permanently frozen.

(b) Little Ice Age in Iceland

In Iceland, also, climatically-associated suffering was apparent. The population fell in numbers, often drastically, as, for example, during 1753–9, when a 25% drop occurred. Indeed the total population only recovered its eleventh-century value during the present century. Summers too cold and wet to dry the hay, winters such as 1695 or 1756, when the island was enclosed by ice for over eight months, and volcanic disasters were undoubtedly responsible for acute difficulties. In 1784, the Danish government actually debated the proposition that Iceland be evacuated and its population resettled in mainland Europe.

(c) Little Ice Age in Scotland and England

In Scotland the climatic margin moved dramatically downhill. Figure 8.9 shows how areas which were formerly viable for agriculture at higher levels were no longer used as the Little Ice Age progressed. Runs of con-

184

Figure 8.9 Climatic deterioration and the agricultural margin in the eastern Southern Uplands of Scotland, 1300–1600. (Source: Parry, 1978)

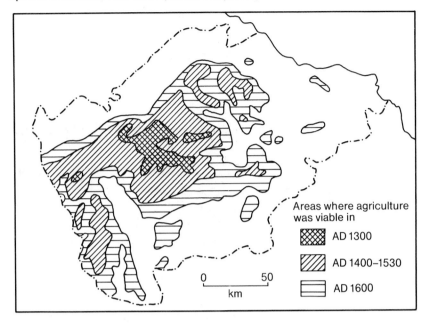

secutive harvest failures in the 1690s and 1780s led to famine years, and in the 'ill years of King William's reign', 1693–1700 (probably the worst years of the Little Ice Age), the oats failed for seven out of eight years in the uplands. The southward spread of the polar sea-ice could be inferred from the appearance of Eskimos in their kayaks around the Orkneys and once in the mouth of the River Don at Aberdeen. Given such bizarre occurrences, it is hardly surprising that the River Thames froze over eleven times in the seventeenth century and that the sea at Marseilles froze in 1595 and 1638. Such events obviously implied great reductions in the growing season throughout Europe, perhaps by as much as two months. Even in England, removed from the worst ravages of the Little Ice Age, rural depopulation (apparent before the arrival of the Black Death), a drop in life expectancy of ten years, and a rise in the average age of marriage were noticeable.

As with all climatic fluctuations, there is a tendency to stress the events falling only on one side of the mean. Thus it is important also to note that even in the depth of the Little Ice Age hot, dry summers and mild winters did occur. Their frequency, however, was substantially less than in the present century.

5. Twentieth-century climatic trends

(a) Warming trend

The recovery from the Little Ice Age is well recorded by instrumental observations. These show a strong warming trend, most apparent in the higher latitudes of the northern hemisphere. This is particularly marked

185

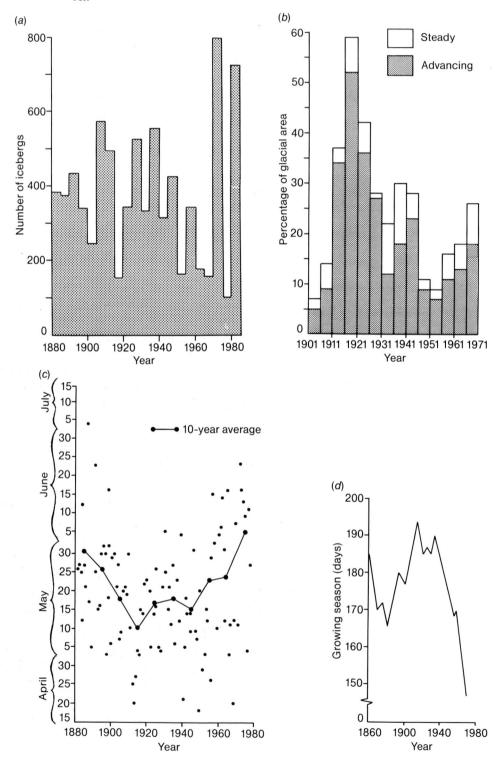

Figure 8.10 Twentieth-century climatic indicators (a) number of icebergs passing south of 48°N (east of Newfoundland), 1880–1985 (five-year average); (b) glacier movements in the western Alps, 1901–71 (five-year average); (c) first summer day in the Netherlands, 1880–1980 (date on which $T_{max} \geqslant 25°C$ at De Bilt); (d) growing season in Toledo in the US corn belt

during the first half of the present century, with a reversal occurring in recent decades. Figure 2.15 shows a global rise of about 0.5°C occurring during the period 1910–40, sufficient to have clearly detectable effects. Some of the consequences of the relatively benign climates of the 1920s, 1930s and 1940s included a recession of many of the Alpine and Arctic glaciers, a diminution of 10–20% in Arctic sea-ice coverage, changes in agricultural practice and changes in fish and animal distributions. A selection of changes is shown in Figure 8.10.

The planetary circulation appears to have been particularly vigorous during this period, transporting warmth to the high latitudes most effectively. Active depressions pushed further polewards, bringing moisture and winter warmth deep into continental interiors. Only in the Americas, where a more vigorous westerly circulation enhances the rain-shadow effect of the Western Cordillera, was rainfall less reliable. The Dust Bowl of the 1930s occurred during one of the most reliable periods for the Asian monsoon and for Sahelian rainfall in Africa. In Britain the growing season lengthened by two weeks, and in the Arctic the ice-free coal-shipping season in Spitsbergen extended from three months in 1920 to seven months in 1939.

(b) Mid-century cooling

Recent decades have seen a reversal of this warming trend, although cooling is as yet confined to the northern hemisphere. The 1970s and 1980s, in particular, have been characterised by more variability in climate than at any time since the Little Ice Age. Examination of Figure 8.11 hints at the role played by circulation changes in this. A weakening of the westerly flow is particularly striking since 1950. This has been compensated for by an increased incidence of other circulation types, most notably those symptomatic of anticyclonic conditions. Such blocking is associated with extremes of weather of various types, depending on the location of the blocking anticyclone. Major droughts (Western Europe 1975, 1976 and 1984; USSR 1972), wet springs (British Isles 1983), hot summers

Figure 8.11 Changing frequencies of the westerly circulation. (Source: Lamb, 1982)

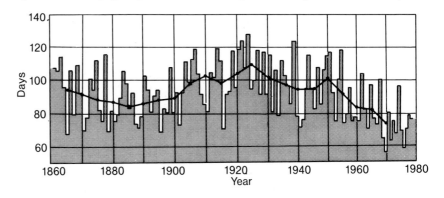

●── 10-year mean, plotted at 5-year intervals

(British Isles 1976 and 1983) and record-breaking cold winter spells (USA 1976/7, Western Europe 1981/2) are all indicative of an increased tendency towards a low zonal index circulation (see Chapter 3). In the tropics, the failure of the equatorial rain belt to extend polewards as far as previously was responsible for the Sahelian drought, and the Asian monsoon has become less reliable, with potentially serious consequences.

Increased variability in climate was particularly disadvantageous for the hybrid strains of wheat and rice introduced in the 1960s, and whose high yields promised alleviation of the food–population problems of the Third World. Some concern exists that this performance, however, is very dependent on high inputs of fertilisers and is rather sensitive to unfavourable climatic conditions. The oil-price rises of the 1970s and the increased variability of climate has hindered the realisation of the full potential of these 'miracle' crops in many areas.

Our continuing susceptibility to climatic variation can be succinctly inferred from Table 8.1. Despite increased technological input, a small perturbation in climate is responsible for a serious diminution in crop productivity – an important lesson regarding the significance of the climatic system.

Table 8.1 Agricultural impact of twentieth-century cooling in Iceland

	Hay yields (tonnes/ha)	Fertiliser inputs (tonnes/ha)	Mean temperature (°C)
1950s	4.33	2.83	7.65
mid-1960s	3.22	4.81	6.83

Source: Goudie, 1977

ASSIGNMENTS

1. Write an essay on the relationship between past climate and human activity.

2. Suggest the chief characteristics which a reconstruction of the atmospheric circulation would show during (a) the medieval warm period and (b) the Little Ice Age.

3. The following data are indicative of the performance of the Indian monsoon from 1947 to 1975. The data show the percentage area of India with insufficient summer rainfall.

1947	1948	1949	1950	1951	1952	1953	1954	1955	1956
0	10	9	8	32	29	10	3	6	0

1957	1958	1959	1960	1961	1962	1963	1964	1965	1966
26	8	20	15	0	22	8	0	60	55

1967	1968	1969	1970	1971	1972	1973	1974	1975
5	45	20	5	20	50	3	56	8

Graph the data and comment on changes in the reliability of the monsoon rains over the period.

D. Causes of Climatic Change

1. The climatic system

The climatic variations described in the preceding section are not random. Rather, they reflect changes – major and minor – over various time-scales, in the workings of what may be labelled the climatic system. Figure 8.12 shows a schematic representation of this climatic system and enables possible reasons for climatic variability to be suggested. Clearly, the state of the climate at a particular time is a function of two types of control. First, the external controls, or 'boundary conditions', consist essentially of the amount of heat energy input to the system from the sun, or escaping from it in the form of outgoing radiation. Second, and more difficult to disentangle, are internal controls involving the component parts of the earth–atmosphere system in storing and distributing this heat input. Complex feedback loops in these categories, for example between the atmosphere and oceans, make the climatic system a difficult one to comprehend, and hence render the causes of climatic change as yet imperfectly understood.

Figure 8.12   Conceptual representation of the climatic system

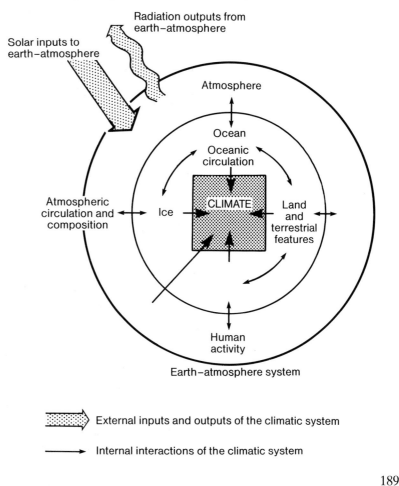

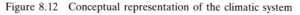 External inputs and outputs of the climatic system

——→   Internal interactions of the climatic system

2. Extra-terrestrial causes

(a) Solar variability

Of primary importance to the climatic system is the impact of solar energy, which drives it. Even a 1% change in this input would cause a corresponding change of about 1°C in global temperature. A reduction of only a few per cent would thus be sufficient to induce a return to glacial climates. Traditionally, however, the sun has been considered a relatively stable star, whose energy output varies only by fractions of 1%, hence the use of the term *solar constant*. The validity of this view has been challenged in recent years, as new research suggests a potential for more variability than was previously envisaged.

One solar characteristic which does exhibit variability is the rise and fall of sunspot activity. These are complicated magnetic phenomena that block the flow of heat from the sun's interior to its surface. Although they are indicative of cooler areas on the surface of the sun (4300°C as opposed to 5700°C), compensating heat output from the areas adjacent to them often means that total solar output is relatively unchanged, or even increased. Furthermore, observations of their number show a fairly reliable eleven-year cycle (Figure 8.13). This cycle parallels solar activity in other forms, such as solar flares and changes in the sun's magnetic field. Superimposed on this eleven-year cycle are others at 22, 80 and 205 years, and it was natural that links were sought between solar indices, as measured by the sunspot cycle, and other phenomena. Inconclusive results were obtained in the majority of cases, and in others good correlations appear coincidental. For example, during the 1960s and early 1970s the rise and fall of Beatle music matched the sunspot cycle, as did the lengths of ladies' hemlines at the Paris fashion shows! More serious attempts include correlations with the growing season in south-west England (Figure 8.14) and with drought years in the mid-western USA. Most convincing of all, though, are the longer-term correlations. Short sunspot cycles of high activity characterised the medieval and twentieth-century warm periods, whereas, long, weakly-formed cycles correspond to cold periods in the climatic record. During the depths of the Little Ice Age hardly any sunspots at all were observed. From 1645 to 1715, the total number counted was less than in some individual years in recent times. Overall, however, the proposition maintaining that such flickers in the solar furnace have significant effects on the workings of the climatic system remains at present not proven.

Figure 8.13 Sunspot numbers, 1700–1980

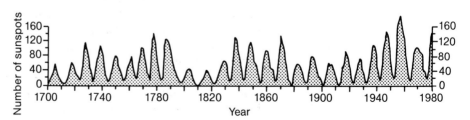

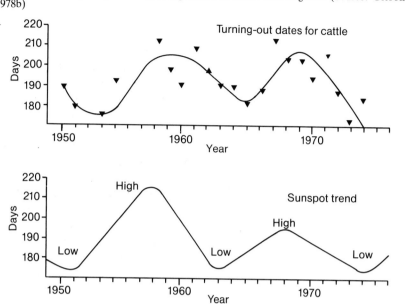

Figure 8.14 Sunspot cycle and growing season in south-west England. (Source: Gribbin, 1978b)

(b) Orbital characteristics of the earth

A dramatic breakthrough in our understanding of the causes of long-term climatic change, especially the oscillation from glacial to interglacial, has been achieved in recent years. This is based on an understanding of the role of cyclical variations in the earth's orbit around the sun. The gravity of the sun, moon and other planets produces three different types of variation which, when combined, exert an important climatic influence (Figure 8.15).

First, the earth's orbit is not an exact circle, but an ellipse; therefore the earth is nearer the sun at certain times of the year. At present, the time of closest approach is on 2/3 January, when the earth as a whole receives about 7% more solar radiation than on 5/6 July, when it is furthest away. Over a period of 96 000 years this orbit changes shape, becoming almost circular, before stretching back to an ellipse. At its most elliptical, seasonal change in the amount of solar radiation reaching the earth approaches 30%. This changing eccentricity, it should be noted, has no effect on total annual energy income, merely on its seasonal and geographical distribution.

Second, the times of year at which closest (*perihelion*) and furthest (*aphelion*) distances occur also vary cyclically. Although at present perihelion is in January, 10 500 years ago it was in June. A rotation of the elliptical orbit over a period of 21 000 years occurs. This is known as the *precession of the equinoxes*, and again seasonal and geographical changes in energy income result.

Third, the tilt of the earth's axis varies, with a period of 40 000 years, between 21.8° and 24.4°. Currently the tilt is 23.5°, and is decreasing by about 0.0001° per annum. The greater the tilt, the more the polar regions

191

Figure 8.15   Cyclical variations in the earth's orbit

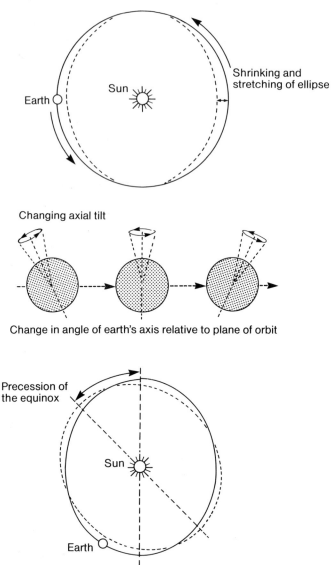

lean towards the sun in summer, and consequently the warmer their summers are. This is especially true of the land-dominated high latitudes of the northern hemisphere. When the tilt is smallest, summer warmth is reduced and conditions favourable for glaciers to expand occur.

Calculation of these three aspects, particularly as to when they are in and out of phase with each other, have been made. These show when relatively high or low receipts of radiation at various latitudes must have occurred during former time periods, and when matched with known interglacial and glacial episodes, a good correspondence is obtained. These controls are now firmly established as the 'pacemakers' of the ice ages.

Two puzzles remained. First, the orbital characteristics described should induce glaciation in only one hemisphere at a time. Yet the evidence suggests that glaciations are synchronous in both hemispheres. The answer appears to be that different requirements exist for the growth of ice-sheets in both hemispheres. In the northern hemisphere, cool summers are crucial so that the previous winter's snowfall, lying around on the land surrounding the Arctic Ocean, fails to melt. On the other hand, cool summers are not so important in the southern hemisphere, where winter snowfall occurs predominantly on a water surface. Instead, it is a cold winter, to freeze more ice from the surrounding sea, that best enables the polar ice to advance. Cool northern summers and cold southern winters occur at the same time in the astronomical cycles, and hence glacial episodes in both hemispheres occur roughly in phase. Of course this only applies if the north polar area is surrounded by land and the south polar area by ocean, as at present.

Second, although astronomical considerations may be acceptable as the 'pacemakers', the trigger that initiates change is less obvious. Astronomical considerations have presumably been acting for the entire span of geological time, yet glaciations are features especially characteristic of the last 2 million years. Other considerations may also be involved. For example, the relatively rapid elevation of large areas of land high into the cold atmosphere, with consequent disruption of the upper circulation, may well have been a trigger mechanism for the present series of glaciations. The thrusting upwards of young fold mountains such as the Rockies, Andes, Himalayas by the Alpine mountain-building episode may have been just such a trigger.

From examination of these two problems it is clear that external influences on the climatic system, although they provide an important framework for directing long-term climatic change, are not by themselves initiators of it. Rather, their influence is modified by the geographical characteristics of the surface, that is, by controls internal to the climatic system.

3. Terrestrial causes

(a) *Changes in atmospheric transparency*

(i) *Vulcanicity and climatic change.*   It will be recalled from Chapter 2 that only about half of the radiation received at the top of the atmosphere actually reaches the surface, due to absorption and reflection by the atmosphere. Changing the composition of this atmospheric sun shield will obviously have climatic effects at the surface. Such changes might alter atmospheric transparency for incoming or outgoing radiation. One way of modifying incoming solar radiation is by the accumulation of a dust or aerosol screen in the atmosphere, either derived from anthropogenic sources (the 'human volcano' described in Chapter 2) or from volcanic sources. The latter are as yet by far the major contributors to particulate pollution in the higher levels of the atmosphere, particularly in the stratosphere. When Mt St Helens erupted in 1980, 2 km^3 of material was blown into the atmosphere and, in geological terms, this was not an exceptional eruption. When debris such as this is injected into the stratosphere (on average the bulk of it locates at altitudes of 18–20 km), it is beyond the reach of rain, which normally scavenges particulates most effectively from the troposphere. A stratospheric veil is created as the upper westerlies diffuse the dust cloud throughout the hemisphere. Residence time for small-diameter particles and sulphuric acid droplets may be several years, and during this time they act with diminishing effect to block the downward passage of solar radiation. While the stratosphere may warm slightly, the surface below cools. In 1883, following the disintegration of the volcanic island of Krakatoa, reductions of incoming solar radiation in excess of 10% were recorded at many of the world's observatories. Some 18 km^3 of material was hurled into the atmosphere. More recently, one of the most significant explosive eruptions of the present century occurred in April 1982. With an output of 500 megatonnes of dust, ash and gas, El Chicon in the Yucatan peninsula was slightly greater in output than Mt St Helens and much more climatically significant. Spectacular sunsets occurred over the northern hemisphere as a global veil formed, and a cooling of the northern hemisphere by 0.3–0.5°C was anticipated.

(ii) *Volcanic records and former climates.*   Historically, there appears to be a link between climate and volcanic activity. The warm first half of the present century was accompanied by a lull in activity, broken by Mt Agung in 1963 and with several eruptions during the more recent cooling period. Examination of the yearly layers of ice in the Greenland ice-sheet enables the sulphuric acid content of each layer to be measured as a surrogate index for past volcanic activity. A good match with temperature trends is apparent (Figure 8.16). The Little Ice Age, for example, stands out as a time of increased acidity, whereas the medieval warm period shows the opposite. Further refinement of techniques, particularly considering the location of the volcano and its sulphuric acid emission, may well reveal an important role for vulcanicity in short and medium-term climatic modelling.

Figure 8.16   A record of the amount of acid deposited in the annual layers of ice accumulation in the Greenland ice-sheets (acidity increases downwards in diagram) and temperature trends in the northern hemisphere. (Source: Lamb, 1982)

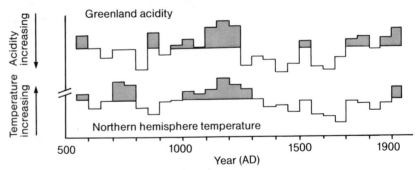

(iii) *Anthropogenic dust loading.*   Changes in atmospheric transparency due to the rapidly increasing human impact on the atmosphere are already detectable. The restriction of outgoing long-wave radiation by increased

194

carbon dioxide concentration, the penetration of more harmful ultraviolet radiation consequent on the damage inflicted by industrial chemicals to the ozone layer, the impact of greatly increased dust loadings in the troposphere – these are just some of the human attacks on the radiation balance involving changing the opacity of the atmosphere. They have already been discussed and the climatic scenarios associated with them outlined in Chapter 2. Whereas in the past humans may, to some extent, have been at the mercy of the climatic system, for the future the converse may well be the case.

(b) Changes in terrestrial geography

(i) *Plate tectonics.* Changes at the surface of the earth can also be hypothesised as causal mechanisms for climatic change. Over very long time-scales, the positions and configurations of the continents change, affecting both local and global climatic patterns. For example, the presence of coal deposits in high-latitude locations such as Spitsbergen is indicative of the fact that tropical swamp-forest conditions existed there during the Carboniferous period of geological time. Deposits of a similar age exist in low-latitude locations such as southern Africa and South America. In these areas, however, the deposits are not coal, but a form of petrified glacial deposit known as *tillite*. Evidently, although present-day low-latitude locations enjoy hot, moist climates, a glacial climate prevailed in part of the present-day tropics. It was partly to resolve such anomalies created by climatically indicative deposits that the *continental-drift hypothesis* was formulated. The coal swamps of eastern North America, Europe and the western USSR developed when these regions enjoyed equatorial locations 300–310 million years ago and when the southern continents lay close to the South Pole. Since lithospheric plates move only at rates of 2–8 cm per annum, however, continental drift cannot readily be invoked to explain climatic variability on scales of less than about 10 million years.

(ii) *Orogenesis.* Mountain-building epochs have also been proposed as a causal mechanism for long-term climatic change. Since temperature falls 6.5°C for every 1000 m rise in altitude, considerable changes may accompany the uplift of areas over tens of thousands of years. As hinted earlier, the Alpine orogenesis may have influenced the pre-existing pattern of preferred wave location and depression/anticyclone formation. It may thus have had a role to play as a forcing function for glaciation within the context of the astronomical cycles previously described.

(iii) *Ocean–atmosphere interaction.* On shorter time-scales of a few years or even several months, the complexities of surface–atmosphere interaction in the form of feedbacks render causes difficult to isolate. This can best be observed with respect to ocean–atmosphere interaction, which is of considerable importance to the climatic system because of the ocean's function as a heat store. As much heat is stored in the upper 3 m of the ocean as in the entire atmosphere, making the ocean a buffer against the more extreme fluctuations of the atmosphere. However, anom-

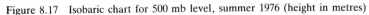

Figure 8.17   Isobaric chart for 500 mb level, summer 1976 (height in metres)

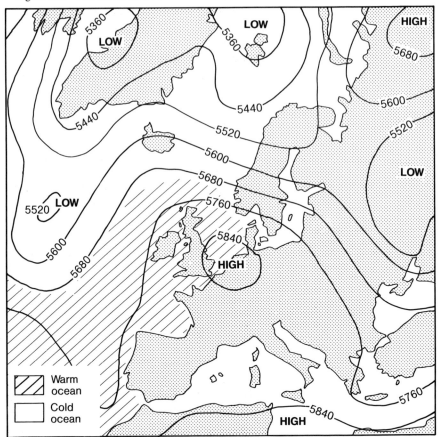

alies of sea-surface temperature can also encourage a particular type of atmospheric circulation to prevail over an extended length of time. This is exemplified in Figure 8.17, which shows the 500 mb chart and the sea-surface temperatures that existed during the West European drought in 1976. The warm weather appears to have led to a warming of the shallow ocean waters around Britain and Ireland, in contrast to the cooler water in mid-Atlantic. The thermal gradient between these two water bodies then appears to have influenced the location of the polar front, displacing it to the north-west of Europe and thus steering depressions along a track to the north-west and enabling the upper air ridge shown in the 500 mb chart to persist. In turn, the warm pool of air over Europe enabled further warming of the ocean beneath, perpetuating the blocking situation.

(iv)   *Land-use/albedo changes*. Surface feedback effects may also be attributable to land-use changes. These were examined in Chapter 2 in the context of albedo changes due to deforestation, overgrazing, soil erosion, etc. and are possibly associated with regional climatic fluctuations such as in the Sahel, or even the global temperature decline of the last three decades. It is particularly imperative to understand such feedback mechanisms, because of the need to anticipate the climatic impact of schemes designed deliberately to change the climate of areas. The diversion of Siberian rivers to the arid areas of central Asia, the damming of the Bering Strait, cloud seeding, hurricane modification – these are just some of the proposed or already operational projects which may or may not have adverse effects on the entire climatic system.

ASSIGNMENTS

1. *List and explain each of the three orbital variables in the astronomical theory of climatic change.*
2. *With reference to Figure 8.16 and the histogram of polar ice in the vicinity of the Icelandic coast (constructed in assignment 1 at the end of section B), deduce which periods during the last thousand years suggest a link between vulcanicity and temperature change.*
3. *Explain, using an example, what is meant by climatic feedback mechanisms.*
4. *Make a case for the proposition that over the next thirty years northern hemisphere temperatures will (a) fall by 0.3°C and (b) rise by 0.3°C.*

E.   The Climatic System: Transitive or Intransitive?

The extent to which the climatic system responds smoothly to external influences is a matter for conjecture. One view is that climate represents an equilibrium response to these influences and readily responds to changes in them. This is known as a *transitive response*. An *intransitive response* would be where the system resists change dictated from outside and tends to remain in the same mode.

Which of these is more appropriate for the climatic system? Perhaps a compromise between the two is most appropriate. Sluggish to respond initially to external influences, once some critical threshold is breached the system moves to a new equilibrium. This is an important finding, if it is correct. First, it implies that more than one climatic state may exist for a particular set of external influences, such as glacial and interglacial modes. Second, however, it has serious repercussions in terms of human impact on climate. If humans breach the unknown threshold and trigger a switch to a new climatic mode, then even eliminating their assault on the atmosphere completely might be insufficient to achieve a reversal.

Key Ideas

A.   Change as the Norm

1. Climate is not fixed, but is inherently variable.
2. Climatic variability will characterise the future as it has the past, and must be planned for.

B. Reconstructing Past Climates

1. Instrumental observations are only of use in identifying recent climatic fluctuations.
2. Proxy records of climate can be derived using such sources as documentary, tree-ring, pollen and isotope ratio evidence.
3. Increasing climatic variability is apparent on longer time-scales.

C. The Climatic Record

1. An alternation between glacial and interglacial climates has occurred often over the past 2 million years. Temperature falls of 10°C accompanied glaciation, as did dislocation of the oceanic and atmospheric circulation.
2. Since the post-glacial optimum, some 5000–7000 years ago, global cooling has occurred.
3. Climatic changes may have had important influences on the geographical spread and cultural development of early peoples.
4. Benign climates from the ninth to the thirteenth century were accompanied by an increase in exploration and settlement by the maritime peoples of north-west Europe, especially the Vikings.
5. From the thirteenth to the nineteenth century, a cooling of the world climate, known as the Little Ice Age, caused harvest failures and hardship in climatically marginal areas.
6. The twentieth century experienced abnormal warmth and reliable rainfall in its earlier decades, although it has turned cooler and more variable after about 1940.
7. Recent increases in climatic variability are potentially detrimental to world food supplies, which are based on high-yielding strains of grain.

D. Causes of Climatic Change

1. Climatic change may occur (i) through external change in the input or output of heat energy or (ii) by internal changes in energy storage or distribution.
2. Long-term changes are strongly linked to orbital characteristics of the earth, which affect the spatial and seasonal receipt of solar energy.
3. Volcanic and anthropogenic dust loading of the atmosphere may be associated with cooling spells on time-scales of a few years.
4. Complex feedback mechanisms involving changes of the surface albedo or ocean heat storage are also operative on short and medium time-scales.

E. The Climatic System: Transitive or Intransitive?

1. The responsiveness of the climatic system to change-inducing factors is not clear. Possible thresholds may exist at which rapid changes in climate may occur.

Additional Activities

1. Figure 8.18(*a*) shows the average frost-free period in the Canadian prairies from 1931 to 1960, and Figure 8.18(*b*) shows the changes that would accompany a drop in mean temperature of 1°C. Wheat takes about 90 days to mature. What conclusions can be drawn about the following:

Figure 8.18 Growing season in the Canadian prairies: (*a*) average frost-free period, 1931– 60; (*b*) changes accompanying a drop in mean temperature of 1°C

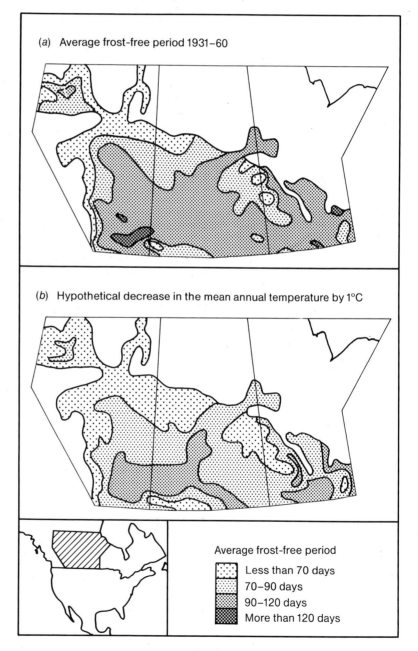

(a) the extent of spatial changes that result from a small temperature change;

(b) the intensity with which climatic variation is experienced in marginal and non-marginal areas;

(c) the sensitivity of harvests to climatic factors as opposed to technological factors?

2. Table 8.2 lists world grain reserves from 1960 to 1985.

(a) Describe the trend that is apparent.

(b) Comment on (i) why the beneficial effects of the Green Revolution appear to decline during the 1970s; (ii) what other factors may be involved.

(c) What significance will climatic fluctuations have on the food–population–energy equation of future years?

Table 8.2 World wheat and coarse grain reserves

Season	End-of season stock (million tonnes)	Reserves as percentage of year's requirements	Season	End-of-season stock (million tonnes)	Reserves as percentage of year's requirements
1960/1	222	26	1973/4	147	12
1961/2	212	24	1974/5	131	11
1962/3	195	21	1975/6	138	11
1963/4	198	21	1976/7	193	15
1964/5	184	19	1977/8	191	14
1965/6	178	18	1978/9	227	16
1966/7	167	17	1979/80	195	14
1967/8	188	19	1980/1	162	14
1968/9	219	21	1981/2	198	17
1969/70	206	19	1982/3	235	19
1970/1	166	15	1983/4	171	14
1971/2	183	16	1984/5	184	14
1972/3	142	12			

Sources: 1960–80 – Lamb, 1982; 1980–5 – *Grains and Oilseeds Review* (August 1984)

Glossary

Adiabatic lapse rate   The change of temperature of an air parcel rising (cooling) or falling (warming) adiabatically. In dry or unsaturated air it is 10°C/km; in moist or saturated air it is variable, but on average about 6°C/km.

Adiabatic process   The process of temperature change as a result of a pressure change, with no exchange of heat or energy with the external environment.

Advection   The horizontal movement of air across a land or sea surface.

Aerosols   Solid particles suspended in the air. They include dust, salt particles, products of combustion, etc.

Air mass   An extensive volume of air possessing uniform physical characteristics at similar heights such as temperature and moisture.

Albedo   The ratio of energy reflected by a surface to the energy incident or falling on that surface.

Anticyclone   A high-pressure system characterised by subsiding air. Downward wind flow is clockwise in the northern hemisphere and anti-clockwise in the southern hemisphere.

Blocking anticyclone   A cellular pattern of high pressure in the mid-latitude zonal circulation belt which diverts or prevents the normal east–west motion of depressions.

Centrifugal force   A force that is exerted on a mass in circular flow and directed outward from the centre. The force is equal and opposite to the centripetal force. The balance of these two forces allows for uniform circular motion.

Climate   'Average' weather over fairly long intervals of time, usually greater than one year and normally about thirty years.

Condensation   The transformation of water vapour to liquid water.

Condensation nuclei   Microscopic particles in the atmosphere that act as a focus or stimulant for cloud-drop growth.

Conduction   Heat transfer through a substance from point to point by means of the movement of adjacent molecular motions.

Convection   The transfer of heat by the actual movement of the heated substance, such as air or water. In meteorology, convection also means vertical transport through density imbalance, transporting mass, water vapour, and aerosols as well as heat (see *advection*).

Convergence   The process where air flow contracts or 'piles up' because of both speed and direction changes. A useful analogy is water flowing through the neck of a funnel.

Coriolis force   The effect of the earth's rotation on air flow. In the northern hemisphere, the Coriolis force deflects air to the right and to the left in the southern hemisphere.

Cyclones (*depressions*)   Weather systems characterised by low pressure and rising air flows. Wind circulation is anticlockwise in the northern hemisphere and clockwise in the southern hemisphere.

Divergence   The process whereby the air flow expands because of changes in speed and direction. A good analogy is water flowing out of the neck of a funnel.

Electromagnetic energy   Energy transferred in the form of disturbances in electric or magnetic fields. One theory suggests that energy is transferred as wave disturbances; the other theory states that energy is transported as particles.

Evaporation   The transformation of liquid water to water vapour.

Front   A boundary separating two air masses such as warm, moist air and cold, dry air. If the cold air pushes into the region of warm air, a cold front occurs and if the warm air advances relative to the cold, a warm front occurs.

Geostrophic wind   Horizontal wind flow as a result of the balance of the pressure-gradient force and the Coriolis force.

Gradient wind   Curved, horizontal wind flow, as a result of the balance of the pressure-gradient force, the Coriolis force and the centrifugal force. Wind flows across the isobars.

Inversion   Temperature increase with increasing height; a negative lapse rate.

Isobar   A line connecting points of equal pressure.

Isotherm   A line connecting points of equal temperature.

Jet stream   A band of high winds usually found in the upper troposphere. Wind speeds can exceed 90 m/s (200 mph).

Latent heat   The amount of energy needed to accomplish a phase change. Latent heat of fusion is the amount of energy required to melt ice, and at 0°C is about 80 cal/g. The latent heat of vapourisation is the amount of energy needed to evaporate liquid water. It is equivalent to about 600 cal/g at 0°C. The latent heat of sublimation is the energy needed to carry out the change from solid (ice) to gas (vapour). It is the sum of the latent heats of fusion and vapourisation, i.e. about 680 cal/g at 0°C. When water freezes, condenses or changes from a gas to a solid, 80 cal/g, 600 cal/g and 680 cal/g are released to the environment, respectively.

Net radiation   The difference between the absorbed and the emitted radiation.

Normal (*environmental*) *lapse rate*   The average change (decrease) of temperature with height. In the troposphere it is 6.5°C/km.

Occluded front   The merging of two fronts, as when a cold front overtakes a warm front.

Polar front   A boundary that separates polar air masses from tropical air masses.

Pressure gradient   The change of pressure with distance. When measured from high to low values perpendicular to the isobars, it is the greatest change in the shortest distance.

Prevailing wind   The most frequent wind direction in a given time period, e.g. day, month, year, etc.

Radiation   This is the transmission of energy by electromagnetic waves, which may be propagated through a substance or through a vacuum at the speed of light. Electromagnetic radiation is divided into various classes on the basis of wavelengths; these are, in order of increasing wavelength: gamma radiation, X rays, ultraviolet radiation, visible (light) radiation, infra-red radiation and radio waves.

Radiosonde   An instrument that measures temperature, pressure and humidity of the upper air. It is carried aloft on a balloon and transmits its measurements to a ground-based receiver via radio signals.

Relative humidity   The ratio of the amount of moisture in the air to the maximum amount of moisture that the air can hold.

Saturation   The condition air reaches when it contains the most water vapour it is capable of holding.

Sensible heat transfer   The transfer of heat by conduction and convection.

Solar radiation   Short-wave electromagnetic energy from the sun.

Source region   Extensive areas where air remains in the same place long enough to acquire the characteristics of an air mass.

Specific heat   The amount of energy required to raise the temperature of one gram of a substance to one degree Celsius.

Specific humidity   The mass of water vapour in a unit mass of moist air.

Stable air   Air is stable when an air parcel sinks or rises to its original position, when the force that initially moved it is no longer operating.

Stratosphere   The atmospheric layer immediately above the troposphere. It extends from 11 km to 45 km above the earth's surface, and temperatures increase with height in this layer.

Sublimation   The process of a solid being transformed directly to a gas or vice versa.

Subsidence   Sinking of air which tends as a result to warm and dry. Inversions usually form in sinking air.

Supercooled water   Liquid water at temperatures below 0°C.

Terrestrial radiation   Long-wave (infra-red) radiation emitted by the earth.

Troposphere   The atmospheric layer closest to the earth's surface. It has an average thickness of 11 km and temperatures on average decrease with altitude.

Turbulence   Atmospheric motion that shows irregular and random motion.

Unstable air   Air parcel that moves away (rises or falls) from its location when the initial force exerted on it is removed.

Weather   The total effect of meteorological elements such as pressure, wind, temperature, humidity, precipitation and cloud cover at any particular place at any particular time.

References and Further Reading

Chapter 1

Neiburger, M., Edinger, J. G. and Bonner, W. O. (1982) *Understanding Our Atmospheric Environment*, W. H. Freeman, San Francisco, pp. 32–4.

Strahler, A. N. and Strahler A. H. (1979) *Elements of Physical Geography*, Wiley, New York, pp. 29–34.

White, I. D., Mottershead, D. N. and Harrison, S. J. (1984) *Environmental Systems*, George Allen & Unwin, London, chs. 1–2.

Chapter 2

Barney, G. O. (1980) *Global 2000 Report to the President of the US*, Vol. 1, Summary Report, Pergamon, New York, pp. 78–91.

Barry, R. G. and Chorley, R. J. (1982) *Atmosphere, Weather and Climate*, Methuen, London, pp. 1–52.

Chandler, T. J. (1981) *Modern Meteorology and Climatology*, Nelson, London, pp. 10–13.

Commission of the European Communities (1979) *The European Solar Radiation Atlas. Vol.1 Global Radiation on Horizontal Surfaces*, W. Grosschen-Verlag.

Gregory, K. J. and Walling, D. E. (eds.) (1979) *Man and Environmental Processes*, Dawson Westview, Kent, pp. 11–22.

Gribbin, J. (1982) *Future Weather: Carbon Dioxide, Climate and the Greenhouse Effect*, Penguin, Harmondsworth.

Jäger, J. (1983) *Climate and Energy Systems: a review of their interactions*, Wiley, New York.

Matthews, W. H., Kellogg, W. W. and Robinson, G. D. (1971) *Man's Impact on the Climate*, MIT Press, Cambridge, Mass., pp. 156–75.

Miller, G. T. (1979) *Living in the Environment*, Wadsworth, London, E21–E31.

Neiburger, M., Edinger, J. G. and Bonner, W. D. (1982) *Understanding our Atmospheric Environment*, W. H. Freeman, San Francisco, pp. 51–91 and 358–78.

Oke, T. R. (1978) *Boundary Layer Climates*, Methuen, London, pp. 31–4.

Riehl, H. (1978) *Introduction to the Atmosphere*, McGraw-Hill, New York, pp. 26–53.

Sagan, C., Toon, O. B. and Pollack, J. B. (1979) Anthropogenic albedo changes and the earth's climate, *Science*, vol. 206(4425), pp. 1363–8.

Sellers, W. D. (1965) *Physical Climatology*, University of Chicago Press, Chicago and London.

Trewartha, G. T. and Horn, L. H. (1980) *An Introduction to Climate*, McGraw-Hill, New York, pp. 8–40.

Chapter 3

Barry, R. G. and Chorley, R. J. (1982) *Atmosphere, Weather and Climate*, Methuen, London.

Battan, L. (1979) *Essentials of Meteorology*, Prentice-Hall, New Jersey.

Bowen, D. (1972) *A Concise Physical Geography*, Hulton, Oxford.

Brimacombe, C. A. (1981) *Atlas of Meteosat Imagery*, European Space Agency, Paris.

Donn, W. L. (1975) *Meteorology*, McGraw-Hill, New York.

Gaskell, T. F. and Morris, M. (1979) *World Climate*, Thames & Hudson, London.

Knapp, B. J. (1981) *Practical Foundations of Physical Geography*, George Allen & Unwin, London.

Lutgens, F. L. and Tarbuck, E. J. (1979) *The Atmosphere*, Prentice-Hall, New Jersey.

Oliver, J. E. (1979) *Physical Geography, Principles and Applications*, Duxbury, California.

Riley, D. and Spolton, L. (1981) *World Weather and Climate*, Cambridge University Press, Cambridge.

Taylor, F. W., Elson, L. S., McCleese, D. J. and Diner, D. J. (1981) Comparative Aspects of Venus and Terrestrial Meteorology, *Weather*, vol. 36(2), pp. 34–40.

Trewartha, G. T. and Horn, L. H. (1980) *An Introduction to Climate*, McGraw-Hill, New York.

Chapter 4

Atkinson, B. W. (1981) *Meso-scale Atmospheric Circulations*, Academic Press, London, pp. 80–92.

Barry, R. G. (1981) *Mountain Weather and Climate*, Methuen, London, pp. 256–64.

Brimacombe, C. A. (1981) *Atlas of Meteosat Imagery*, European Space Agency, Paris.

Chorley, R. J. (ed.) (1969) *Water, Earth and Man*, Methuen, London, pp. 11–29.

Ernst, J. A. (1976) SMS–1 Night-time infrared imagery of low level mountain waves, *Monthly Weather Review*, vol. 104, pp. 207–9.

Flohn, H. (1969) *Climate and Weather*, Weidenfeld & Nicolson, London, pp. 42–78.

HMSO (1982) *Cloud Types for Observers*, Meteorological Office, London.

Ludlam, F. H. (1978) The forms and patterns of cumulus, *Weather*, vol. 33(2), pp. 54–63.

Riehl, H. (1978) *Introduction to the Atmosphere*, McGraw-Hill, New York, pp. 73–104

Scorer, R. (1972) *Clouds of the World: a complete colour encyclopedia*, David & Charles, Newton Abbot.

Spiegel, H. J. and Gruber, A. (1983) *From Weather Vanes to Satellites: an introduction to meteorology*, Wiley, New York, pp. 49–56.

Trewartha, G. T. and Horn, L. H. (1980) *An Introduction to Climate*, McGraw-Hill, New York.

Wheeler, D. (1984) The July 1983 'heatwave' in north-west England, *Weather*, vol. 39(6), pp. 178–81.

Chapter 5

Atkinson, B. W. (1977) *Urban Effects of Precipitation: an investigation of London's influence on the severe storm in August 1975*. Department of Geography, Queen Mary College, University of London. Occasional Paper No. 8.

Atkinson, B. W. and Smithson, P. A. (1976) Precipitation. In T. J. Chandler and S. Gregory (eds.), *The Climate of the British Isles*, Longman, London, pp. 129–82.

Barrett, E. C. and Martin, D. W. (1981) *The Use of Satellite Data in Rainfall Monitoring*, Academic Press, London, pp. 3–16.

Barry, R. G. (1981) *Mountain Weather and Climate*, Methuen, London, pp. 180–93.

Breuer, G. (1980) *Weather Modification: prospects and problems*, Cambridge University Press, Cambridge.

Chandler, T. J. and Gregory, S. (eds.) (1976) *The Climate of the British Isles*, Longman, London.

HMSO (1973) *British Rainfall – 1967*, Meteorological Office, London.

Landsberg, H. (1974) Inadvertent atmospheric modification through urbanisation. In W. N. Hess (ed.), *Weather and Climate Modification*, Wiley, New York, pp. 726–63.

Landsberg, H. (1981) *The Urban Climate*, Vol. 28 in the International Geophysics Series, Academic Press, New York.

Woodley, W. L., Griffith, C. G., Griffin, J. and Augustine, J. (1978) Satellite rain estimation in the Big Thompson and Johnstown flash floods. Preprint, Conference on Flash Floods: Hydro-meteorological Aspects, Los Angeles 2–5 May 1978.

Chapter 6

Barke, M. and O' Hare, G. (1984) *The Third World*, Oliver & Boyd, Edinburgh.

Chandler, T. J. (1981) *Modern Meteorology and Climatology*, Nelson, London, pp. 58–67.

Doornkamp, J. C. and Gregory, K. (1980) *Atlas of Drought in Britain (1975–1976)*, Institute of British Geographers, London.

Flohn, H. (1969) *Climate and Weather*, Weidenfeld & Nicolson, London, pp. 156–93.

HMSO (1983) *Tables of Temperature, Relative Humidity, Precipitation and Sunshine for the World. Part IV: Africa, the Atlantic Ocean south of 35°N and the Indian Ocean*, Meteorological Office, London.

Hutchinson, P. and Sam, J. A. (1984) The unusual start of the wet season in the Gambia – 1982, *Weather*, vol. 39(1), pp. 24–8.

Jackson, I. J. (1977) *Climate, Water and Agriculture in the Tropics*, Longman, London, pp. 33–103.

Lockwood, J. C. (1974) *World Climatology; an environmental approach*, Arnold, London, pp. 28–30.

Manshard, W. (1979) *Tropical Agriculture*, Longman, London, pp. 19–24.

Nieuwolt, S. (1977) *Tropical Climatology*, Wiley, New York, pp. 102–26.

Petterssen, S. (1969) *Introduction to Meteorology*, McGraw-Hill, New York, pp. 261–75.

Riehl, H. (1978) *Introduction to the Atmosphere*, McGraw-Hill, New York, pp. 358–76.

Roy, M. G., Hough, M. N. and Starr, J. R. (1978) Some agricultural effects of the drought of 1975–76 in the United Kingdom, *Weather*, vol. 33(2), pp. 64–74.

Shakesby, R. A. and Trilsbach, A. (1982) Irrigation is the desert's secret agent, *Geographical Magazine*, vol. LIV (2), pp. 77–83.

Trewartha, G. T. and Horn, L. H. (1980) *An Introduction to Climate*, McGraw-Hill, New York, pp. 192–213.

Whittow, J. (1980) *Disasters: an anatomy of environmental hazards*, Allen Lane, London, pp. 248–309.

Chapter 7

Brimbacombe, C. A. (1981) *Atlas of Meteosat Imagery*, European Space Agency, Paris.

Fotheringham, R. R. (1979) *The Earth's Atmosphere Viewed from Space*, University of Dundee, Dundee.

Oliver, J. (1973) *Climate and Man's Environment*, Wiley, New York.

Oliver, J. E. (1979) *Physical Geography, Principles and Applications*, Duxbury, California.

Riley, D. and Spolton, L. (1981) *World Weather and Climate*, Cambridge University Press, Cambridge.

Trewartha, G. T. and Horn, L. H. (1980) *An Introduction to Climate*, McGraw-Hill, New York.

University of Dundee (1984) Satellite Pictures – 1 January 1984, *Weather*, vol. 39(3), pp. 80–1.

University of Dundee (1984) Satellite Pictures – 20 February 1984, *Weather*, vol. 39(4), pp. 126–7.

Chapter 8

Flohn, H. (1974) *Climate and Weather*, World University Library, London.

Fritts, H. (1976) *Tree Rings and Climate*, Academic Press, London.

Goudie, A. S. (1977) *Environmental Change*, Oxford University Press, Oxford.

Gribbin, J. (1978a) *Climatic Change*, Cambridge University Press, Cambridge.

Gribbin, J. (1978b) *The Climatic Threat*, Fontana, Glasgow.

Gribbin, J. (1982) *Future Weather: Carbon Dioxide, Climate and the Greenhouse Effect*, Penguin, Harmondsworth.

Imbrie, J. and Imbrie, K. (1979) *Ice Ages: Solving the Mystery*, Macmillan Press, New York.

Lamb, H. H. (1982) *Climate, History and the Modern World*, Methuen, London.

Lockwood, J. (1979) *Causes of Climate*, Arnold, London.

Mitchell, J. M., et al. (1975) *Climate, Climatic Change and Water Supply*, National Academy of Sciences, Washington, DC.

Oliver, J. (1973) *Climate and Man's Environment*, Wiley, New York.

Parry, M. (1978) *Climatic Change and Agricultural Settlement*, Archon Books, Folkestone.

Roberts, W. and Lansford, H. (1979) *The Climate Mandate*, Freeman, San Francisco.

Skinner, B. J. (1981) *Climates Past and Present*, Kaufmann, Los Altos.

Index